T0214618

Lecture Notes
in Business Information Processing   390

Series Editors

Wil van der Aalst ⓘ
RWTH Aachen University, Aachen, Germany
John Mylopoulos ⓘ
University of Trento, Trento, Italy
Michael Rosemann ⓘ
Queensland University of Technology, Brisbane, QLD, Australia
Michael J. Shaw
University of Illinois, Urbana-Champaign, IL, USA
Clemens Szyperski
Microsoft Research, Redmond, WA, USA

More information about this series at http://www.springer.com/series/7911

Ralf-Detlef Kutsche · Esteban Zimányi (Eds.)

Big Data Management and Analytics

9th European Summer School, eBISS 2019
Berlin, Germany, June 30 – July 5, 2019
Revised Selected Papers

Springer

Editors
Ralf-Detlef Kutsche
Technische Universtät Berlin
Berlin, Germany

Esteban Zimányi ⓘ
Université libre de Bruxelles
Brussels, Belgium

ISSN 1865-1348   ISSN 1865-1356   (electronic)
Lecture Notes in Business Information Processing
ISBN 978-3-030-61626-7   ISBN 978-3-030-61627-4   (eBook)
https://doi.org/10.1007/978-3-030-61627-4

This Springer imprint is published by the registered company Springer Nature Switzerland AG
The registered company address is: Gewerbestrasse 11, 6330 Cham, Switzerland

Preface

The 9th European Big Data Management and Analytics Summer School (eBISS 2019[1]) took place in Berlin, Germany, in July 2019. Tutorials were given by renowned experts and covered advanced aspects of analytics and big data. This volume contains the lecture notes of the summer school.

The first chapter is devoted to actionable conformance checking. In the context of business processes, conformance checking aims at comparing a process model with an event log of the same process in order to assess whether the actual execution of a business process conforms to the model and vice versa. Although conformance checking has been receiving increasing attention in the last years, making the output of a conformance checking process actionable is still a real challenge. This chapter provides an introductory overview of the main techniques of the conformance checking field. In order to make it actionable, simple Python code snippets are provided to illustrate how an organization can start a conformance checking project on its own data. The chapter also provides pointers to open-source scripting libraries that can be used to make conformance checking and process mining actionable.

The second chapter provides an introduction to text analytics. It starts by presenting sources of textual data and the main challenges in text analysis. The chapter then surveys the various steps and methods involved in a typical processing pipeline. Since the steps to be realized heavily depend on the analytical task that is to be achieved, it is therefore necessary to identify the problem at hand and align the process accordingly. The chapter provides illustrative examples in each of the steps of the process and concludes by describing potential applications of text analytics, including sentiment analysis and automatic generation of content.

The third chapter is devoted to automated machine learning. Nowadays, machine learning techniques and algorithms are employed in almost every application domain to extract valuable knowledge from the massive amounts of data produced every day in our digital world. However, building a high-quality machine learning model is an iterative, complex, and time-consuming process that requires knowledge and experience. Given the continuous increase of the amount of digital data produced, it has been acknowledged that the number of data scientists cannot scale to address these challenges. The chapter gives an overview of the state-of-the-art tools and frameworks that have been proposed for tackling the challenges of machine learning automation. It concludes by discussing some research directions and open challenges required to achieve the vision and goals of automated machine learning.

The fourth chapter addresses the problem of determining how travel time can be computed from GPS data. The volume of GPS data collected from moving vehicles has increased significantly over the last years. Nowadays, it is possible to analyze the traffic on most of the road networks without installing roadside equipment. The chapter

---

[1] http://cs.ulb.ac.be/conferences/ebiss2019/.

presents a generic data model for travel time prediction that has a global scope and is applicable when GPS data and a road network graph is present. It defines several weather classes (dry, fog, rain, and snow) and shows their impact on travel time in various road categories (motorway, secondary, tertiary, and residential). The paper also analyzes other weather characteristics such as outside temperature and wind as well as regional differences. These results are presented in the context of a large-scale nationwide study performed in Denmark, where GPS data collected from 10,560 vehicles over five years is integrated with OpenStreetMap data and detailed weather information from the NOAA.

The last chapter introduces the Laplacian matrix as an efficient tool for addressing various tasks in machine learning. Many machine learning problems can be expressed by means of a graph with nodes representing training samples and edges representing the relationship between samples in terms of similarity, temporal proximity, or label information. As graphs can be represented by matrices, the chapter advocates the use of a Laplacian matrix, which allows us to assign each node a value that varies only slightly between strongly connected nodes and more between distant nodes. Such an assignment can be used to extract a useful feature representation, find a good embedding of data in a low dimensional space, or perform clustering on the original samples. The chapter starts by introducing the Laplacian matrix and then presents several algorithms designed around it for data visualization and feature extraction.

In addition to the lectures corresponding to the chapters described above, there were four additional lectures, as follows:

- Ralf-Detlef Kutsche from Technische Universität Berlin, Germany: Science Methodology
- Begüm Demir from Technische Universität Berlin, Germany: Deep Earth Query, Advances in Remote Sensing Image Characterization and Indexing from Massive Archives
- Aymen Cherif from Eura Nova, Belgium: Deep Learning, Current Applications and Future Trends
- Albert Bifet from Télécom ParisTech, France: Machine Learning for Data Streams

These lectures have no associated chapter in this volume.

As for the previous editions, eBISS joined forces with the Erasmus Mundus IT4BI-DC consortium and hosted its doctoral colloquium aiming at community building and promoting a corporate spirit among PhD candidates, advisors, and researchers of different organizations. The corresponding two sessions, each organized in two parallel tracks, included the following presentations:

- Judith Awiti, Evolving ETL workflows in a big data environment
- Jam Jahanzeb Behan, Statistical multidimensional data modeling based on Linked Open Data
- Moditha Hewasinghage, Physical design in document stores
- Mohsin Iqbal, Spatio-textual analytics
- Suela Isaj, Multi-source spatial entity linkage
- Nusrat Jahan Lisa, Database operations on top of complex system design
- Rediana Koci, A data-driven approach to prescribe Web API evolution

- Subba Lawan, Bitmap indexing for big data
- Shumet Tadesse Nigatu, Semi-automatic generation of data intensive APIs
- Olga Rybnytska, Prescriptive analytics for physical systems models

We would like to thank the attendants of the summer school for their active participation, as well as the speakers and their co-authors for the high quality of their contribution in a constant evolving and highly competitive domain. Finally, we would like to thank the external reviewers for their careful evaluation of the chapters.

June 2020   Ralf-Detlef Kutsche
                                                             Esteban Zimányi

Organization

The 9th European Big Data Management and Analytics Summer School (eBISS 2019) was organized by the Technische Universität Berlin, Germany, and the Department of Computer and Decision Engineering (CoDE) of the Université libre de Bruxelles, Belgium.

Program Committee

Alberto Abelló	Universitat Politécnica de Catalunya, BarcelonaTech, Spain
Ralf-Detlef Kutsche	Technische Universität Berlin, Germany
Boudewijn van Dongen	Technische Universiteit Eindhoven, The Netherlands
Nacéra Bennacer	CentraleSupélec, France
Esteban Zimányi	Université libre de Bruxelles, Belgium

External Referees

Judith Awiti	Université libre de Bruxelles, Belgium
Oscar Romero	Universitat Politècnica de Catalunya, BarcelonaTech, Spain
Mahmoud Sakr	Université libre de Bruxelles, Belgium
Alejandro Vaisman	Instituto Tecnológica de Buenos Aires, Argentina
Stijn Vansummeren	Université libre de Bruxelles, Belgium
Robert Wrembel	Poznan University of Technology, Poland
Jianqiu Xu	Nanjing University of Aeronautics and Astronautics, China

Sponsorship and Support

Education, Audiovisual and Culture Executive Agency (EACEA)

Contents

Actionable Conformance Checking: From Intuitions to Code

Josep Carmona[1]([⊠]), Matthias Weidlich[2], and Boudewijn van Dongen[3]

[1] Universitat Politècnica de Cataluna, Barcelona, Spain
jcarmona@cs.upc.edu
[2] Humboldt University of Berlin, Berlin, Germany
matthias.weidlich@hu-berling.de
[3] Eindhoven University of Technology, Eindhoven, The Netherlands
B.F.v.Dongen@tue.nl

Abstract. Conformance checking is receiving increasing attention in the last years. This is due to several reasons, that can be summarized into two: the explosion of digital information that talks about processes, and the need to use this data in order to monitor and improve processes in organizations. Naturally, conformance checking addresses this by providing techniques capable of relating modeled and recorded process information. This paper overviews in a very accessible way the main techniques and feedback of the conformance checking field. Moreover, in order to make it actionable, code snippets are provided so that an organization can start a conformance checking project on its own data.

Keywords: Conformance checking · Process mining · Business process management · BPMN · Petri nets · Event logs · Alignments

1 Introduction

Nowadays organizations are facing a digital transformation, that primarily requires active use of the tons of data available as a result of their operation. As processes are the main focus for the management of an organization, exposing processes to the data available helps to assess the alignment between observed and modeled behavior. When modeled and observed behavior are aligned, then one can be sure that the reality and the models describing it agree. In contrast, an organization may need to react in case of finding deviations between observed and modeled behavior. Conformance checking techniques [1] tackle this fundamental problem: to analytically asses the adequacy of a process model in representing the traces in an event log, extracting the deviations in case they exist. Due to the potential existence of regulations, guidelines, frauds and errors, conformance checking is becoming an essential element for an organization to prove the adherence to a desired behavior.

Conformance checking is a crucial dimension in process mining [2]: by relating modelled and observed behavior, process models that have either been discovered

© Springer Nature Switzerland AG 2020
R.-D. Kutsche and E. Zimányi (Eds.): eBISS 2019, LNBIP 390, pp. 1–24, 2020.
https://doi.org/10.1007/978-3-030-61627-4_1

or manually created, can be confronted with event data. On its core, conformance checking relies on the fundamental problem of identifying, among the set of runs of a process model (which can be infinite), the run that mostly resembles an observed trace.

In general, conformance checking has been applied to very different domains, including healthcare, banking, finance, transportation, manufacturing among others. The reader can see detailed use cases of all these fields in the web of the *IEEE Task Force on Process Mining*: https://www.tf-pm.org.

In this paper we aim at providing a gentle introduction to the conformance checking field, by describing its main techniques. Furthermore, we show code snippets illustrating some of the conformance checking techniques presented in this paper. The code snippets provided in this paper and related data is available in https://github.com/matthiasweidlich/conf_tutorial/.

2   Related Work

The field of conformance checking is relatively new. The definition of the area and a proposal of initial algorithms was presented in the scope of Anne Rozinat's PhD thesis at the TU/e [3] and corresponding publications [4–7]. Important notions arise from this work, like *fitness* or *appropriateness* between a process model and log. Also, important algorithms result from this work, including the techniques to evaluate fitness based on the replay of the traces and the missing/remaining/produced/consumed tokens. Also in the scope of the TU/e, the seminal work under the PhD thesis of Arya Adriansyah is crucial for formalizing the notion of *alignments* [8]. Several applications of alignments are explored in the related publications, like performance analysis [9,10], high-level deviations [11], privacy analysis of user behaviour [12], and alignment-based precision metrics [13].

Another work that has been important for conformance checking is the log conformance analysis presented in the scope of Matthias Weidlich's PhD thesis [14]. The thesis introduces the concept of *behavioural profiles*, as a tailored abstraction for processes that allows comparing recorded and modelled behaviour.

3   Process Models and Event Logs

Process models and event logs represent different conceptualizations of processes. When describing a process, a process model provides an abstraction, capturing some of the process' activities by means of *tasks*. A specific instance of a process, i.e., a case, then corresponds to a sequence of task executions, denoted *run*. In contrast, event logs store the executions of a certain process in a organization. In the remainder of this section we informally introduce these two conceptualizations with the help of a real-life example.

A process model that describes how a loan application is handled is illustrated in Fig. 1. This model is captured in the Business Process Model and Notation (BPMN). In BPMN, tasks are represented by rectangles; instantaneous events

are visualised by circles (in Fig. 1 they start or end the process); and execution dependencies are modelled by control flow arcs and diamond-shaped nodes, called gateways. The semantics of such a gateway determines the exact behaviour of a process, e.g., whether incoming arcs are synchronised (AND-gateway with a 'plus' symbol) or not (XOR-gateway with a 'cross' symbol); or whether outgoing arcs are enabled concurrently (AND-gateway) or mutually exclusive to each other (XOR-gateway). A run of the model (a sequence from start to end that agrees with the aforementioned semantics) is $\langle As, Aa, Fa, Sso, Ro, Ao, Aaa, Af \rangle$.

According to this model, a submitted application is either accepted or rejected, based on the aforementioned rules to check plausibility of the applicant's data. An accepted application is finalised by a worker, in parallel with the offer process. For each application, an offer is selected and sent to the customer. The customer reviews the offer and sends it back. If the offer is accepted, the process continues with the approval of the application and the activation of the loan. If the customer declines the offer, the application is also declined and the process ends. However, the customer can also request a new offer, in which case the offer is cancelled and a new offer is sent to the customer.

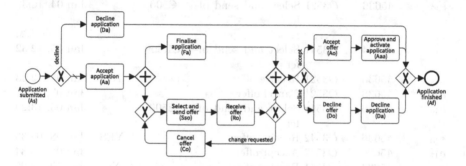

Fig. 1. Example process model of a loan application process in BPMN.

The recording of a single execution of an activity of a process in a information system is called an *event*. Typically, events are performed in a certain *context*, such as for example a specific loan application. This context is commonly given by the case as part of which an activity was executed. The notion of a case, therefore, binds together events, thereby allowing us to track the evolution of a case over time. The events related to a single case are called a *trace*.

The notion of a trace is fundamental for *event logs*. In essence, an event log is a collection of traces, each trace comprising events that can be sorted by their occurrence time. Consider for example our loan application process of Fig. 1. Table 1 shows an excerpt of such event log. The application with id $A5634$ is accepted by the system on January 1st at 12:32 and the customer asks for a €2,000 loan. On January 3rd the application is finalised and two days later, an offer is made to the customer for a €1,500 loan. The offer is received back on January 10th and the customers did not sign it, nor did they indicate they want

Table 1. Example of a log of the loan application process.

Event	Application	Offer	Activity	Amount	Signed	Timestamp
...	...	...	...	...	...	...
e_{13}	$A5634$		Application submitted	€2,000		Jan 01, 12:31
e_{14}	$A5634$		Accept application	€2,000		Jan 01, 12:32
e_{15}	$A5635$		Application submitted	€5,000		Jan 02, 04:31
e_{16}	$A5635$		Accept application	€5,000		Jan 02, 04:32
e_{17}	$A5636$		Application submitted	€200		Jan 03, 06:59
e_{18}	$A5636$		Accept application	€200		Jan 03, 07:00
...	...	...	...	...	...	...
e_{22}	$A5634$		Finalise application			Jan 03, 09:00
e_{23}	$A5636$		Finalise application			Jan 03, 09:01
e_{24}	$A5635$		Decline application			Jan 03, 09:02
e_{25}	$A5635$		Decline application			Jan 03, 09:03
...	...	...	...	...	...	...
e_{30}	$A5636$	$O3521$	Select and send offer	€500		Jan 04, 16:32
...	...	...	...	...	...	...
e_{37}	$A5634$	$O3541$	Select and send offer	€1,500		Jan 05, 12:32
e_{38}	$A5636$	$O3521$	Receive offer		NO	Jan 05, 12:33
e_{38}	$A5636$	$O3521$	Cancel offer			Jan 05, 12:34
e_{39}	$A5636$	$O3542$	Select and send offer	€500		Jan 05, 13:29
e_{40}	$A5636$	$O3542$	Receive offer		YES	Jan 08, 08:33
e_{41}	$A5636$	$O3542$	Accept offer			Jan 08, 16:34
e_{42}	$A5634$	$O3541$	Receive offer		NO	Jan 10, 10:00
...	...	...	...	...	...	...
e_{54}	$A5634$	$O3541$	Decline offer			Jan 10, 10:04
...	...	...	...	...	...	...
e_{64}	$A5634$		Decline application			Jan 10, 10:05
e_{65}	$A5634$		Application finished			Jan 10, 10:06
e_{66}	$A5636$		Approve and activate application			Jan 10, 10:07
e_{67}	$A5636$		Application finished			Jan 10, 10:08
...	...	...	...	...	...	...

any changes. Therefore, a few minutes later, the offer is declined, which is also done for the application as a whole.

4   Conformance Checking

4.1   Quality Dimensions to Relate Process Models and Event Logs

By relating observed and modeled behavior, an organization can get insights on the execution of their processes with respect to the expectations as described in the models. If both process model M and event log L are considered as languages, their relation can be used to measure how good is a process model in describing the behavior recorded in an event log.

Hence, confronting M and L can help into understanding the complicate relation between modeled and recorded behavior. We now provide two visions of this relation, that represent two alternative perspectives: *fitness* and *precision*.

Fitness measures the ability of a model to explain the recorded execution of a process as recorded in an event log (see the example of Fig. 2 for an example of fitting behavior). It is the main measure to assess whether a model is well-suited to explain the recorded behaviour. To explain a certain trace, the process model is queried to assess its ability in replaying the trace, taking into account the control flow logic expressed in the model.

In general, fitness is the fraction of the behaviour of the log that is also allowed by the model. It can be expressed as follows.

$$fitness = \frac{|L \cap M|}{|L|} \tag{1}$$

Let us have a look at this fraction in more detail by examining the extreme cases. Fitness is 1, if the entire behaviour that we see in the log L is covered by the model M. Conversely, fitness is 0, if no behaviour in the log L is captured by the model M. In the Sect. 4.2 we will describe three different algorithms deriving artefacts that can be used to evaluate fitness.

We define a trace to be either *fitting* (it corresponds to a run of the model) or *non-fitting* (there is some deviation with respect to all runs of the model). For instance, the trace corresponding case $A5634$ in our running example is fitting, since there is a model run that perfectly reproduces this case, as shown in Fig. 2. In contrast, Fig. 3 shows the information for a trace that does not contain the event to signal that the application has been finalised (Fa).

Precision is the counterpart of fitness. It can be calculated by looking at the fraction of the model behaviour that is covered in the log.

$$precision = \frac{|L \cap M|}{|M|} \tag{2}$$

We see that precision shares the numerator in the fraction with fitness from (1). This implies that if we have a log and a model with no shared behaviour, fitness is zero, and by definition also precision is zero. However, the denominator is replaced with the amount of modelled behaviour.

In summary, for the two main metrics reported above, algorithms that can assess the relation between log and model need to be considered. In the next section, we describe the three main algorithmic perspectives to accomplish this task.

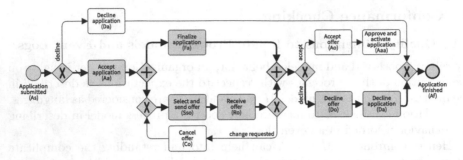

Fig. 2. Loan application process model with highlighted path corresponding to the fitting trace of case $A5634$ from the event log of Table 1.

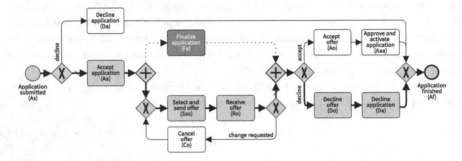

Fig. 3. Loan application process model with highlighted path corresponding to one trace, which does not include an event to signal that the application has been finalised (Fa). In magenta we show that the task (Fa) has not been observed but it is required to reach the final state of the process model.

4.2   Computing Conformance Checking Artefacts

The relation between a trace observed in the event log, and a process model, is described as a *conformance checking artefact*. In this section we will introduce three possible conformance checking artefacts, overviewed in Fig. 4. The reader is refered to [1] for a detailed explanation of the contents of this section.

Rule Checking. The basic idea of rule-based conformance checking is to exploit rules that are satisfied by all the runs of a process model as the basis for analysis. Such rules define a set of constraints that are imposed by the process model. The verification of these constraints with respect to the traces of an event log, therefore, enables the identification of conformance issues.

Considering the running example of our loan application process as depicted in Fig. 1, rules derived from the process model include:

R1: An application can be accepted (Aa) at most once.

R2: An accepted application (Aa), that must have been submitted (As) earlier, and eventually an offer needs to be selected and sent (Sso) for it.

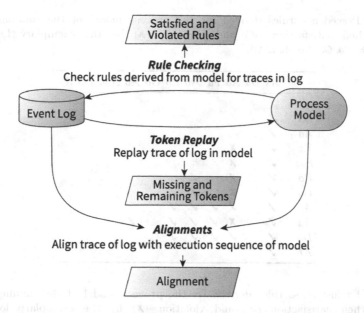

Fig. 4. General approaches to conformance checking and resulting conformance artefacts (from [1]).

R3: An application must never be finalised (*Fa*), if the respective offer has been
declined (*Do*) already.

R4: An offer is either accepted (*Ao*) or declined (*Do*), but cannot be both
accepted and declined.

A careful inspection of each one of the rules above would reveal that they
are different in nature: rule R1 is an example of *cardinality rule*, which defines
an upper and lower bound for the number of executions of an activity. Rule
R2 contains a *precedence rule*, which establishes that the execution of a certain
task is preceded by at least on execution of another task. Rule R3 establishes an
ordering rule, whereas rule R4 represents an *exclusiveness rule*. Tables 2 and 3
show examples of cardinality and exclusiveness rules, respectively, for the running example and two log traces.

By assessing to what extent the traces of a log satisfy the rules derived
from a process model, rule-based conformance checking focuses on the fitness
dimension, i.e., the ability of the model to explain the recorded behaviour. Traces
are fitting, if they satisfy the rules, or non-fitting if that is not the case. Let R_M
be a predefined set of rules. Fitness can be defined according[1] to R_M:

$$\text{fitness}(L, M) = \frac{|\{r \in R_M \mid r \text{ is satisfied by all } t \in L\}|}{|R_M|} \tag{3}$$

---

[1] Notice that this makes fitness to depend on a particular set of rules, which is a
limitation of the rule-based fitness checking.

Table 2. Precedence rules derived for the process model of the running example and their satisfaction (✓) and violation (✗) by the exemplary log trace ⟨As, Sso, Fa, Ro, Co, Ro, Aaa, Af⟩.

	As	Da	Aa	Fa	Sso	Ro	Co	Ao	Aaa	Do	Af
As											
Da	✓										
Aa	✓										
Fa	✓	✗									
Sso	✓	✗									
Ro	✓	✗		✓							
Co	✓	✗		✓	✓						
Ao	✓			✓	✓	✓	✓				
Aaa	✓	✗		✓	✓	✓		✗			
Do	✓			✓	✓	✓	✓				
Af	✓										

Table 3. Exclusiveness rules derived for the process model of the running example and their satisfaction (✓) and violation (✗) by the exemplary log trace ⟨As, Aa, Sso, Ro, Fa, Ao, Do, Da, Af⟩.

	As	Da	Aa	Fa	Sso	Ro	Co	Ao	Aaa	Do	Af
As	✓										
Da		✓						✗	✓		
Aa			✓								
Fa				✓							
Sso											
Ro											
Co											
Ao		✗						✓	✗		
Aaa		✓						✓	✓		
Do								✗	✓	✓	
Af											✓

As the reader may already have grasped, the dimension of precision is not targeted by rule-checking.

Token Replay. Intuitively, this technique replays each trace of the event log in the process model by executing tasks according to the order of the respective events. By observing the states[2] of the process model during the replay, one can

---

[2] A state of a BPMN model is a distribution of tokens over the control flow arcs. A task is enabled in a state if its incoming control flow arc is assigned a token by the respective distribution. If it executes, this token is *consumed*, i.e., no longer assigned to the arc. Moreover, a token is *produced* on the outgoing control flow arc of the task.

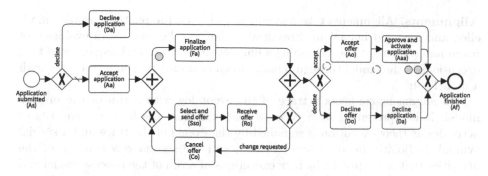

Fig. 5. State reached after replaying the full trace $\langle As, Aa, Sso, Ro, Ao, Aaa, Aaa \rangle$. One can see that there are three remaining tokens (denoted by yellow background), and two missing tokens (denoted by discontinuous red lines). (Color figure online)

determine whether, and to what extent, the trace indeed corresponds to a valid run of the process model.

In essence, token replay postulates that each trace in the event log corresponds to a valid execution sequence of the process model. This is verified by step-wise executing tasks of the process model, according to the order of the respective events in the trace. During this replay, we may observe two cases that hint at non-conformance (see Fig. 5):

(i) the execution of a task requires the consumption of a token on the incoming arc, but the arc is not assigned any token in the current state, i.e., a token is *missing* during replay;

(ii) the execution of a task produces a token at an outgoing arc, but this token is not consumed eventually, i.e., a token is *remaining* after replay.

By exploring whether the replay of a trace yields missing or remaining tokens, replay-based conformance checking mainly focuses on the fitness dimension. That is, the ability of the model to explain the recorded behaviour is the primary concern. Traces are fitting if their replay does not yield any missing or remaining tokens, and non-fitting otherwise:

$$\text{fitness}(L, M) = \frac{1}{2}\left(1 - \frac{\sum_{t \in L} missing(t, M)}{\sum_{t \in L} consumed(t, M)}\right) + \frac{1}{2}\left(1 - \frac{\sum_{t \in L} remaining(t, M)}{\sum_{t \in L} produced(t, M)}\right)$$
$$(4)$$

In contrast to rule checking, precision can be estimated using token replay [15], but unfortunately, the corresponding technique strongly relies on the assumption that traces are fitting; if they are not, then the estimation of precision through token replay can be significantly degraded [13].

Alignments. Alignments take a symmetric view on the relation between modelled and recorded behaviour. Specifically, they can be seen as an evolution of token replay. Instead of establishing a link between a trace and sequences of task executions in the model through replay, alignments directly connect a trace with a model run.

An alignment connects a trace of the event log with a run of the process model. It is represented by a two-row matrix, where the first row consists of activities as their execution is signalled by the events of the trace and a special symbol $\gg$ (jointly denoted by e_i below), and the second row consists of the activities that are captured by task executions of a run of the process model and a special symbol $\gg$ (jointly denoted by a_i):

$$\begin{array}{l|c|c|c|c|} \text{log trace} & e_1 & e_2 & \ldots & e_n \\ \hline \text{model run} & a_1 & a_2 & \ldots & a_m \end{array}$$

Each column in this matrix, a pair (e_i, a_i), is a *move* of the alignment, meaning that an alignment can also be understood as a sequence of moves. There are different types of such moves, each encoding a different situation that can be encountered when comparing modelled and recorded behaviour. We consider three types of moves:

- *Synchronous move*: A step in which the event of the trace and the task in the run correspond to each other. Synchronous moves denote the expected situation that the recorded events in the trace are in line with the tasks of a run of the process model. In the above model, a synchronous move means that it holds $e_i = a_i$ and $e_i \neq \gg$ (and thus $a_i \neq \gg$).
- *Model move*: When a task should have been executed according to the model, but there is no related event in the trace, we refer to this situation as a model move. As such, the move represents a deviation between the trace and the run of the process model in the sense that the execution of an activity has been skipped. In the above model, a model move is denoted by a pair (e_i, a_i) with $e_i = \gg$ and $a_i \neq \gg$.
- *Log move*: When an event in the trace indicates that an activity has been executed, even though it should not have been executed according to the model, the alignment contains a log move. Being the counterpart of a model move, a log move also represents a deviation in the sense of a superfluous execution of an activity. A log move is denoted by a pair (e_i, a_i) with $e_i \neq \gg$ and $a_i = \gg$.

Alignments are constructed only from these three types of moves (see an in-depth explanation on this in [1]).

For instance, let us use the running example (see Fig. 1) and the trace $\langle As, Aa, Sso, Ro, Ao, Aaa, Aaa \rangle$. A possible alignment with this trace is:

$$\begin{array}{l|c|c|c|c|c|c|c|c|c|} \text{log trace} & As & Aa & Sso & Ro & \gg & Ao & Aaa & Aaa & \gg \\ \hline \text{model run} & As & Aa & Sso & Ro & Fa & Ao & Aaa & \gg & Af \end{array}$$

This alignment comprises six synchronous moves, one log move, $(Aaa, \gg)$, and two model moves, $(\gg, Fa)$ and $(\gg, Af)$. The log move $(Aaa, \gg)$ indicates that the application had been approved and activated, even though this was not expected in the current state of processing (as this had just been done). The model move $(\gg, Fa)$ is the situation of the process model requiring that the application be finalised, which has not been done according to the trace. Furthermore, one can easily extract the original trace by projecting away the special symbol for skipping from the top row. Applying the projection to the bottom row yields the run of the model $(\langle As, Aa, Sso, Ro, Fa, Ao, Aaa, Af \rangle)$.

In general, *optimal alignments*, i.e., alignments with minimal number of move or log moves, are preferred. The alignment shown above is optimal since there is no other alignment with least number of deviations. Computing (optimal) alignments is a hot research topic, which has been addressed in many papers in the last years [8, 16–27]. In this paper, however, we will refrain from describing the state-of-the-art methods for alignment computation, and refer the interested reader to the aforementioned papers, or to [1].

Remarkably, alignments provide a simple means to quantify fitness. Again, this may be done based on the level of an individual trace or the event log as a whole. However, the aggregated cost of log moves and model moves may be a misleading measure, though, as it is not normalised. A common approach, therefore, is to normalise this cost by dividing it by the worst-case cost of a aligning the trace with the given model. Under a uniform assignment of costs to log and model moves, such a worst-case cost originates from an alignment in which each event of the trace T_i relates to a log move, whereas all task executions of a run σ of the model relate to a model move and σ is as short as possible. Since the cost induced by the model moves of a model run depends on its length, the shortest possible model run leading from the initial state to a final state in the model is considered for this purpose.

Realising the above idea, we obtain two ratios that denote the relative share of non-fitness in the alignments of a trace or an event log, respectively. Let M be a model and L an event log. Then, we denote by $cost(t, M)$ the cost of an optimal alignment of a trace $t \in L$ with respect to the model. Furthermore, let $cost(t, \langle \rangle)$ and $cost(\langle \rangle, x)$ be the costs of aligning a trace t with an empty model run, or some run $x \in M$ of the model with an empty trace, respectively. Then, fitness based on alignments is quantified for a trace or an event log:

$$\text{fitness}(L, M) = 1 - \left(\frac{\sum_{t \in L} cost(t, M)}{\sum_{t \in L} \left(cost(t, \langle \rangle) \right) + |L| \times \min_{x \in M} cost(\langle \rangle, x)} \right) \quad (5)$$

A simple precision metric based on alignments is grounded in the general idea of *escaping edges* [15]. To give the intuition, we assume that (i) the event log fits the process model; and (ii) that the process model is deterministic. The former means that we simply exclude non-fitting traces, for which the optimal alignment contains log moves or model moves, from the assessment of the precision of the model. The latter refers to a process model not being able to reach a state, in

which two tasks that capture the same activity of the process are enabled. The model of our running example (see Fig. 1) is deterministic.

For the activity of each event of a trace of the event log, we can determine a state of the process model right before the respective task would be executed. Under the above assumptions, this state is uniquely characterised. What is relevant when assessing precision, is the number of tasks enabled in this state of the process model. Let M be a process model and L an event log, with $t \in L$ as a trace and, overloading notation, $e \in t$ as one of the events of the trace. Then, by $enabled_M(e)$, we denote the number of tasks and, due to determinism of the process model also the number of activities that can be executed in the state right before executing the task corresponding to e.

Similarly, we consider all traces of the log that also contain events related to the activity of event e, say a, and have the same prefix, i.e., events that indicate that the same sequence of activities has been executed before an event signalling the execution of activity a. Then, we determine the number of activities for which events signal the execution directly after this prefix, i.e., the set of activities that have been executed in the same context as the activity a as indicated by event e. Let this number of activities be denoted by $enabled_L(e)$, which, under the above assumptions, is necessarily less than or equal to $enabled_M(e)$. Then, the ratio of both numbers captures the amount of 'escaping edges' that represent modelled behaviour that has not been recorded. As such, precision of log L and M is quantified as follows:

$$\text{precision}(L, M) = \frac{\sum_{t \in L, e \in t} enabled_L(e)}{\sum_{t \in L, e \in t} enabled_M(e)} \tag{6}$$

5   Code Snippets for Conformance Checking

In the previous section an informal introduction to conformance checking has been provided. Concepts like event log, process model, deviation, rule checking, token replay, alignment, fitness and precision should now be familiar to the reader. They are meant to define the complicate relation between modeled and recorded behavior. In this section we take the reader to practice, by introducing simple and intuitive Python code to make most of the aforementioned concepts actionable. Hopefully, the contents of this section can contribute to unleash the application of conformance checking.

5.1   Event Log Exploration

We start by providing examples on how to read an event log, and for extracting different types of information from it. The code provided is a subset of the one available in the repository used for this paper, where several other analyses can be found. The following code reads a log in XES format, the standard format for event logs approved by the IEEE [28] (Fig. 6).

```
 1   import xml.etree.ElementTree as et
 2
 3   def load_xes(file):
 4       log = []
 5
 6       tree = et.parse(file)
 7       data = tree.getroot()
 8
 9       # find all traces
10       traces = data.findall('{http://www.xes-standard.org/}trace')
11
12       for t in traces:
13           trace_id = None
14
15           # get trace id
16           for a in t.findall('{http://www.xes-standard.org/}string'):
17               if a.attrib['key'] == 'concept:name':
18                   trace_id = a.attrib['value']
19
20           events = []
21           for event in t.iter('{http://www.xes-standard.org/}event'):
22
23               e = {'name': None, 'timestamp': None, 'resource': None,
                  ↪  'transition': None}
24
25               for a in event:
26                   e[a.attrib['key'].split(':')[1]] = a.attrib['value']
27
28               events.append(e)
29
30           # add trace to log
31           log.append({'trace_id': trace_id, 'events': events})
32
33       return log
```

Fig. 6. Code for reading an event log.

Once a log is read, one can extract valuable information from traversing it. For instance, the following code shows the length of the shortest and the longest trace in the log.

```
1  log_file = 'conf_tutorial/financial_log.xes'
2  log = load_xes(log_file)
3  max_length = 0
4  min_length = 1000
5
6  for trace in log:
7      if len(trace['events']) > max_length:
8          max_length = len(trace['events'])
9
10     if len(trace['events']) < min_length:
11         min_length = len(trace['events'])
12
13 print('The longest trace contains %s events. The shortest trace: %s
   ↪   events.' %(max_length, min_length))
```

Also, the number of *trace variants*, i.e., number of different traces, of the log can be determined:

```
1  trace_list = []
2
3  for trace in log:
4      events = []
5      for event in trace['events']:
6          events.append(event['name'])
7
8      trace_list.append(tuple(events))
9
10 trace_variants = set(trace_list)
11
12 print('The log contains %s trace variants.' %len(trace_variants))
```

Events in the event log may have several attributes, like a timestamp or a resource. We can use these timestamps to compute the duration of a single trace. The following code returns the shortest and longest duration of all traces in the event log.

```
1  def get_timestamp(input_str: str):
2      """
3      Method to convert a string into a timestamp.
4
5      :param input_str: timestamp as string
6      """
7      timestamp_format = '%Y-%m-%dT%H:%M:%S.%f%z'
8
9      return datetime.strptime(''.join(input_str.rsplit(':', 1)),
       ↪   timestamp_format)
10
11 from datetime import datetime, timedelta
12
13
14 max_duration = timedelta(microseconds=1)
15 min_duration = timedelta(days=10000)
16
17 for trace in log:
18
19     # we only need to consider the first and last event in the
       ↪   trace
20     first_e = trace['events'][0]
21     last_e = trace['events'][-1]
22
23     t0 = get_timestamp(first_e['timestamp'])
24     t1 = get_timestamp(last_e['timestamp'])
25     duration = t1 - t0
26
27     if duration > max_duration:
28         max_duration = duration
29     elif duration < min_duration:
30         min_duration = duration
31
32 print('The shortest process instance took %s; the longest %s'
   ↪   %(min_duration, max_duration))
```

As a final illustration of event log exploration, we focus on another event attribute. In the following code, we output how many different resources are used across the process instances, and the ratio of events that are processed by a resource.

```
1  resources = []
2
3  for trace in log:
4      for event in trace['events']:
5          resources.append(event['resource'])
6
7  print('All process instances use %s different resource in total' %
   ↪  len(set(resources)))
8  no_res = resources.count(None)
9  print('%.2f%% of all events are processed by a resource.'
   ↪   %(((len(resources)-no_res)/len(resources))*100 ))
```

5.2   The Computation of Conformance Checking Artefacts

We now consider how conformance checking artefacts can be computed so that
deviations between modeled and recorded behavior can be obtained.

Process models will be assumed to be defined as Petri nets. In the repository
provided with this paper, a Petri net Python class (denoted PetriNet in the
code) will be used, which contains the standard helper functions to manage it.
We assume the reader to be familiar with Petri nets in this paper (if not, a nice
tutorial can be found in [29]). The following code reads a process model for the
running example, sets the initial state, and finally draws it.

```
1  %run ./conf_tutorial/pn.py
2
3  net = PetriNet()
4  load(net, "./conf_tutorial/financial_log_80_noise.pnml")
5
6  # mark the initial place
7  net.add_marking(1,1)
8  # visualise it
9  draw_petri_net(net)
```

Importantly, mapping events in the event log and tasks in the process model
is an important step so that the conformance checking artefacts can be computed. The following code sets up some helper dictionaries to relate Petri net
transition IDs and activity labels in the event log to each other. Observe that
for the sake of simplicity, an activity label is only assigned to a single transition.
However, multiple transitions may carry a τ label, representing a silent transition
(a transition that does not correspond to any event in the log).

```
1  # helper mappings between ids and labels
2  mapping = net.get_mapping()
3  rev_mapping = {}
4  for k, v in net.get_mapping().items():
5      for k2 in v:
6          rev_mapping[k2] = k
7
8  from pprint import pprint
9  # mapping from labels to LISTS of transitions ids
10 pprint(mapping)
11
12 # mapping from transitions id to label
13 pprint(rev_mapping)
```

The next code illustrates how, given an initial marking, the currently enabled transitions may be identified, how the marking is changed by firing a transition, and how the marking may be adapted to enable a transition.

```
1  print("Initial marking: ", net.get_marking())
2
3  enabled = net.all_enabled_transitions()
4  print("Enabled transitions in initial marking: ",
5       list(map((lambda k: rev_mapping[k]), enabled)))
6
7  # Fire enabled transition (take the first, but there is only one)
8  net.fire_transition(enabled[0])
9  enabled = net.all_enabled_transitions()
10 print("Enabled transitions after firing first transition: ",
11      list(map((lambda k: rev_mapping[k]), enabled)))
12
13 # Check whether the transition with label 'O_CREATED' is enabled
14 # (there is only one transition carrying this label)
15 print("Is transition 'O_CREATED' enabled?",
16
17      net.is_enabled(net.get_mapping()['O_CREATED'][0]))
18 # Enable the transition by changing the marking and adding tokens to
   ↪    the input
19 # places of the transition with label 'O_CREATED'
20 input_places =
   ↪    net.get_input_places(net.get_mapping()['O_CREATED'][0])
21
```

```
22  for p in input_places:
23      net.add_marking(p,1)
24
25  # Again, check whether the transition with label 'O_CREATED' is
    ↪  enabled
26  print("Is transition 'O_CREATED' enabled after tokens have been
    ↪  added to the places in its preset?",
27          net.is_enabled(net.get_mapping()['O_CREATED'][0]))
28
29  # Check whether further transitions have been enabled by adding the
    ↪  token to
30  # the places in the preset of the transition with label 'O_CREATED'
31  enabled = net.all_enabled_transitions()
32  print("Enabled transitions after adapting the marking: ",
33          list(map((lambda k: rev_mapping[k]), enabled)))
34
35  print("Current marking: ", net.get_marking())
```

We are now ready to define and use conformance checking artefacts. We will start with rule checking. Specifically, we consider a cardinality rule that checks a lower and an upper bound for the number of executions of an activity for a particular trace, as well as an ordering rule that checks whether executions of one activity happen only after executions of another activity.

More concretely, we check whether the five most frequent trace variants satisfy the following rules:

1. The application is completed at least once (activity "W_Completeren aanvraag").
2. The application is submitted at most once (activity "A_SUBMITTED").
3. The income lead ("W_Afhandelen leads") is fixed only after the preacceptance ("A_PREACCEPTED"), but never before.

```
1   def check_lower_bound(trace: [], act: str, bound: int) -> bool:
2       count = trace.count(act)
3       return count >= bound
4
5   def check_upper_bound(trace: [], act: str, bound: int) -> bool:
6       count = trace.count(act)
7       return count <= bound
8
9   def check_order_after(trace: [], act_1: str, act_2: str) -> bool:
10      if act_1 not in trace or act_2 not in trace:
```

```
11        return True
12        idx_1 = [i for i, x in enumerate(trace) if x == act_1]
13        idx_2 = [i for i, x in enumerate(trace) if x == act_2]
14        return idx_1[0] >= idx_2[-1]
15
16    # compute the trace variants sorted by frequency
17    trace_variants = {}
18    for trace in log:
19        events = []
20        for event in trace['events']:
21            events.append(event['name'])
22        trace_variants[tuple(events)] =
      ↪   trace_variants.get(tuple(events), 0) + 1
23    trace_variants_sorted_by_freq = sorted(trace_variants.items(),
      ↪   key=lambda kv: kv[1], reverse=True)
24
25    for k in range(5):
26        trace_k = list(trace_variants_sorted_by_freq[k][0])
27        print("Checking trace: %s" % trace_k)
28        print("Application completed at least once? ",
          ↪   check_lower_bound(trace_k, 'W_Completeren aanvraag', 1))
29        print("Application submitted at most once? ",
          ↪   check_upper_bound(trace_k, 'A_SUBMITTED', 1))
30        print("Fixing income lead only after preacceptance? ",
          ↪   check_order_after(trace_k, 'W_Afhandelen leads',
          ↪   'A_PREACCEPTED'))
```

We can also apply token replay on the running example. The following code illustrates how to do token replay for a trace, and how to evaluate fitness for the 30 most frequent variants of the event log.

```
1    def replay_trace(net: PetriNet, trace: []) -> (int, int, int, int):
2        produced = 1
3        consumed = 1
4        missing = 0
5
6        # replay trace, event by event
7        for event in trace:
8            # identify transition, assumption here is that there is only
          ↪   one transition for the label
9            transition = net.get_mapping()[event][0]
10           # check if the transition is enabled
11           if not net.is_enabled(transition):
12               # not enabled, so add a token to all input places that
              ↪   are not marked
13               for p in net.get_input_places(transition):
14                   if net.marking[net.index_of_place(p)] == 0:
```

```
15                          # record the token as missing
16                          missing += 1
17                          net.add_marking(p, 1)
18
19          # record the numbers produced and consumed tokens when
            ↪ firing the transition
20          produced += len(net.get_input_places(transition))
21          consumed += len(net.get_output_places(transition))
22          net.fire_transition(transition)
23
24      # we expect one token left, everything else counts as remaining
25      remaining = sum(net.get_marking()) - 1
26      return produced, consumed, missing, remaining
27
28
29  def fitness(net: PetriNet, log_freq: dict) -> float:
30      sum_prod = 0
31      sum_cons = 0
32      sum_miss = 0
33      sum_rema = 0
34
35      for trace_var, freq in log_freq.items():
36          # keep copy of marking
37          marking = list(net.get_marking())
38          # replay trace
39          replay_values = replay_trace(net, trace_var)
40          sum_prod += log_freq[trace_var] * replay_values[0]
41          sum_cons += log_freq[trace_var] * replay_values[1]
42          sum_miss += log_freq[trace_var] * replay_values[2]
43          sum_rema += log_freq[trace_var] * replay_values[3]
44          # restore marking
45          for k,v in net.places.items():
46              net.add_marking(v, marking[k])
47
48      return 0.5 * (1 - sum_miss / sum_cons) + 0.5 * (1 - sum_rema /
            ↪ sum_prod)
49
50  fitness_value = 0
51  for k in range(30):
52      log_k = {t[0]:t[1] for t in
            ↪ trace_variants_sorted_by_freq[k:k+1]}
53      log_x = {t[0]:t[1] for t in
            ↪ trace_variants_sorted_by_freq[0:k+1]}
54      fitness_value_k = fitness(net, log_k)
55      fitness_value = fitness(net, log_x)
56      print("Fitness value of the single %s-most frequent trace
            ↪ variant: %f" % (k+1, fitness_value_k))
57      print("Fitness value of %s-most frequent trace variants: %f" %
            ↪ (k+1, fitness_value))
```

Finally, we provide code to illustrate how alignments can be also used as conformance checking artefact. The following code illustrates how to use them to show deviations. We will be using the alignment functionality that is contained in the Python class Astar, provided also in the repository of this paper. In the following code snippets, we will use some of the objects computed before, like the Petri net, and the most frequent variants in the event log.

```
1  from pprint import pprint
2  %run ./conf_tutorial/alignment.py
3
4  # select some most frequent traces
5  traces = dict()
6  for k in range(10):
7      traces[k] = list(trace_variants_sorted_by_freq[k][0])
8
9  # capture details on which places denote the start and the end of
   ↪  the process model
10  index_place_start = 0
11  index_place_end = 1
12
13  # run alignment construction
14  a = Astar()
15  alignments = a.Astar_Exe(net, traces, index_place_start,
   ↪  index_place_end, no_of_solutions=1)
16
17  # print the alignments
18  for k,t in traces.items():
19    print('Trace in the log: ', t)
20    print('Optimal alignment: ')
21    pprint(alignments[k][0])
```

And now alignment-based fitness can be reported, as illustrated in the code below:

```
1  def fitness(net: PetriNet, alignments: list, log_freq: list) ->
   ↪  float:
2      # the shortest model run in our example contains seven elements
3      shortest_seq_in_net = 7
4
5      async_moves = 0
6      max_cost = 0
7
8      for k in range(len(alignments)):
9          async_moves += log_freq[k] * len([x for x in alignments[k]
   ↪  if (x[0] == '-' or x[1] == '-')])
10         max_cost += log_freq[k] * (shortest_seq_in_net +
   ↪  len(alignments[k]))
11
12      return round(float(async_moves) / float(max_cost), 3)
```

```
13
14  fitness_value = 0
15  for k in range(len(alignments)):
16      alignments_simple_k = [alignments[k][0]]
17      alignments_simple = [alignments[x][0] for x in range(k+1)]
18      log_freq_k = [trace_variants_sorted_by_freq[k][1]]
19      log_freq = [trace_variants_sorted_by_freq[x][1] for x in
        ↪  range(k+1)]
20      fitness_value_k = fitness(net, alignments_simple_k, log_freq_k)
21      fitness_value = fitness(net, alignments_simple, log_freq)
22      print("Fitness value of the single %s-most frequent trace
        ↪  variant: %f" % (k+1, fitness_value_k))
23      print("Fitness value of %s-most frequent trace variants: %f" %
        ↪  (k+1, fitness_value))
```

6   Concluding Remarks

In this paper we have introduced conformance checking and provided code snippets to make the discipline actionable in practice. The paper focuses in the definition and use of the main conformance checking artefacts, namely rule checking, token replay and alignments, so that a clear insight on the relation between modeled and observed behavior can be obtained from them.

To make it accessible, we have chosen to stay on simple, specially tailored, Python code that is sufficient for the main purpose of this paper. For the reader that became interested, we strongly advice to look for other open-source scripting libraries that can be also used to make conformance checking and process mining actionable: PMLAB [30], BupaR [31], pm4py [32] are some examples.

Acknowledgments. This work has been supported by MINECO and FEDER funds under grant TIN2017-86727-C2-1-R.

References

1. Carmona, J., van Dongen, B.F., Solti, A., Weidlich, M.: Conformance Checking - Relating Processes and Models. Springer, Cham (2018). https://doi.org/10.1007/978-3-319-99414-7
2. van der Aalst, W.M.P.: Process Mining - Data Science in Action, 2nd edn. Springer, Heidelberg (2016). https://doi.org/10.1007/978-3-662-49851-4
3. Rozinat, A.: Process mining conformance and extension. Ph.D thesis, Technische Universiteit Eindhoven (2010)
4. Rozinat, A., van der Aalst, W.M.P.: Conformance testing: measuring the fit and appropriateness of event logs and process models. In: Bussler, C.J., Haller, A. (eds.) BPM 2005. LNCS, vol. 3812, pp. 163–176. Springer, Heidelberg (2006). https://doi.org/10.1007/11678564_15

5. van der Aalst, W.M.P., Dumas, M., Ouyang, C., Rozinat, A., Verbeek, H.M.W.E.: Choreography conformance checking: an approach based on BPEL and Petri nets. In: The Role of Business Processes in Service Oriented Architectures, 16 July–21 July2006 (2006)
6. van der Aalst, W.M.P., Dumas, M., Ouyang, C., Rozinat, A., Verbeek, H.M.W.E.: Conformance checking of service behavior. ACM Trans. Internet Technol. 8(3), 13:1–13:30 (2008)
7. Rozinat, A., van der Aalst, W.M.P.: Conformance checking of processes based on monitoring real behavior. Inf. Syst. 33(1), 64–95 (2008)
8. Adriansyah, A.: Aligning observed and modeled behavior. Ph.D. thesis, Technische Universiteit Eindhoven (2014)
9. Adriansyah, A., Buijs, J.C.A.M.: Mining process performance from event logs. In: La Rosa, M., Soffer, P. (eds.) BPM 2012. LNBIP, vol. 132, pp. 217–218. Springer, Heidelberg (2013). https://doi.org/10.1007/978-3-642-36285-9_23
10. van der Aalst, W.M.P., Adriansyah, A., van Dongen, B.F.: Replaying history on process models for conformance checking and performance analysis. Wiley Interdiscip. Rev. Data Min. Knowl. Discov. 2(2), 182–192 (2012)
11. Adriansyah, A., van Dongen, B.F., Zannone, N.: Controlling break-the-glass through alignment. In: International Conference on Social Computing, Social-Com 2013, SocialCom/PASSAT/BigData/EconCom/BioMedCom 2013, Washington, DC, USA, 8–14 September, 2013, pp. 606–611 (2013)
12. Adriansyah, A., van Dongen, B.F., Zannone, N.: Privacy analysis of user behavior using alignments. Inf. Technol. 55(6), 255–260 (2013)
13. Adriansyah, A., Munoz-Gama, J., Carmona, J., van Dongen, B.F., van der Aalst, W.M.P.: Measuring precision of modeled behavior. Inf. Syst. e-Bus. Manag. 13(1), 37–67 (2015)
14. Weidlich, M.: Behavioural Profiles: A Relational Approach to Behaviour Consistency. Doctoral thesis, Universität Potsdam (2011)
15. Muñoz-Gama, J., Carmona, J.: A fresh look at precision in process conformance. In: Hull, R., Mendling, J., Tai, S. (eds.) BPM 2010. LNCS, vol. 6336, pp. 211–226. Springer, Heidelberg (2010). https://doi.org/10.1007/978-3-642-15618-2_16
16. van Dongen, B.F.: Efficiently computing alignments - using the extended marking equation. In: Weske, M., Montali, M., Weber, I., vom Brocke, J. (eds.) BPM 2018. LNCS, vol. 11080, pp. 197–214. Springer, Cham (2018). https://doi.org/10.1007/978-3-319-98648-7_12
17. Taymouri, F., Carmona, J.: Model and event log reductions to boost the computation of alignments. In: Ceravolo, P., Guetl, C., Rinderle-Ma, S. (eds.) SIMPDA 2016. LNBIP, vol. 307, pp. 1–21. Springer, Cham (2018). https://doi.org/10.1007/978-3-319-74161-1_1
18. de Leoni, M., Marrella, A.: Aligning real process executions and prescriptive process models through automated planning. Expert Syst. Appl. 82, 162–183 (2017)
19. Reißner, D., Conforti, R., Dumas, M., Rosa, M.L., Armas-Cervantes, A.: Scalable conformance checking of business processes. Paper submitted to "International Conference on Business Process Management (BMP 2017)" in Barcelona, Spain, March 2017
20. Leemans, S.J.J., Fahland, D., van der Aalst, W.M.P.: Scalable process discovery and conformance checking. Softw. Syst. Model. 17(2), 599–631 (2016). https://doi.org/10.1007/s10270-016-0545-x
21. García-Bañuelos, L., van Beest, N.R., Dumas, M., Rosa, M.L., Mertens, W.: Complete and interpretable conformance checking of business processes. IEEE Trans. Softw. Eng. 44(3), 262–290 (2018)

22. Taymouri, F., Carmona, J.: A recursive paradigm for aligning observed behavior of large structured process models. In: 14th International Conference of Business Process Management (BPM), Rio de Janeiro, Brazil, 18–22 September (2016)
23. Taymouri, F., Carmona, J.: An evolutionary technique to approximate multiple optimal alignments. In: Weske, M., Montali, M., Weber, I., vom Brocke, J. (eds.) BPM 2018. LNCS, vol. 11080, pp. 215–232. Springer, Cham (2018). https://doi.org/10.1007/978-3-319-98648-7_13
24. van Dongen, B., Carmona, J., Chatain, Th., Taymouri, F.: Aligning modeled and observed behavior: a compromise between computation complexity and quality. In: Dubois, E., Pohl, K. (eds.) CAiSE 2017. LNCS, vol. 10253, pp. 94–109. Springer, Cham (2017). https://doi.org/10.1007/978-3-319-59536-8_7
25. Bloemen, V., van de Pol, J., van der Aalst, W.M.P.: Symbolically aligning observed and modelled behaviour. In: 18th International Conference on Application of Concurrency to System Design, ACSD 2018, Bratislava, Slovakia, 25–29 June 2018, pp. 50–59 (2018)
26. Taymouri, F., Carmona, J.: Structural computation of alignments of business processes over partial orders. In: 19th International Conference on Application of Concurrency to System Design, ACSD 2019, Aachen, Germany 23–28 June 2019, pp. 73–81 (2019)
27. Padró, L., Carmona, J.: Approximate computation of alignments of business processes through relaxation labelling. In: Hildebrandt, T., van Dongen, B.F., Röglinger, M., Mendling, J. (eds.) BPM 2019. LNCS, vol. 11675, pp. 250–267. Springer, Cham (2019). https://doi.org/10.1007/978-3-030-26619-6_17
28. Acampora, G., Vitiello, A., Stefano, B.N.D., van der Aalst, W.M.P., Günther, C.W., Verbeek, E.: IEEE 1849: the XES standard: the second IEEE standard sponsored by IEEE computational intelligence society [society briefs]. IEEE Comp. Int. Mag. **12**(2), 4–8 (2017)
29. Murata, T.: Petri nets: properties, analysis and applications. Proc. IEEE **77**(4), 541–574 (1989)
30. Carmona, J., Solé, M.: PMLAB: an scripting environment for process mining. In: Proceedings of the BPM Demo Sessions 2014 Co-located with the 12th International Conference on Business Process Management (BPM 2014), Eindhoven, The Netherlands, 10 September 2014, p. 16 (2014)
31. Janssenswillen, G., Depaire, B., Swennen, M., Jans, M., Vanhoof, K.: bupaR: enabling reproducible business process analysis. Knowl. Based Syst. **163**, 927–930 (2019)
32. Berti, A., van Zelst, S.J., van der Aalst, W.M.P.: Process mining for python (PM4Py): bridging the gap between process- and data science. CoRR abs/1905.06169 (2019)

Introduction to Text Analytics

Agata Filipowska$^{(\boxtimes)}$ and Dominik Filipiak

Department of Information Systems,
Poznan University of Economics and Business, Poznan, Poland
{agata.filipowska,dominik.filipiak}@ue.poznan.pl
http://www.ue.poznan.pl

Abstract. Data processing regards analysis of various types of data, including numerical data, signals, texts, pictures, videos, etc. This paper focuses on defining and studying various tasks of text analytics following the typical processing pipeline. Sources of textual data are introduced and related challenges are discussed. Along with the process of text analytics, examples are presented to demonstrate how text analytics should be carried out. Finally, potential applications of text analytics are given including sentiment analysis and automatic generation of content.

Keywords: Text analytics · Sentiment analysis · Information extraction · Application scenarios

1 Introduction

Big data analytics concerns processing variety of data from variety of sources. Recent years, companies focus on analysing data coming directly from users/-customers to provide personalised user experience. Examples of such content include customer reviews published online regarding, e.g., products or services purchased. Based on reviews one may not only study the personal attitude of a customer towards a product or a service, but also identify features important from a customer's perspective. Such information may then greatly influence the product development or marketing activities [4].

However, analysis of text faces multiple challenges. Texts coming from blogs, reviews, etc. are an example of a user-generated content, what influences quality (inconsistencies, spelling mistakes, etc.), questions their origin (reviews of products may be written by a producer himself or by a competition) or logical content flow. Also, different languages pose different challenges related to their grammar or inflection as well as technicalities such as encoding [11].

This paper is to provide insights into the process of text analysis and related challenges. Following a tutorial format, it firstly discusses sources of textual data and key aspects of text analysis. Then, along with the process of text analytics, examples are presented to demonstrate how text analytics should be carried out (steps to be presented include tokenization, lemmatization, disambiguation, etc.). Finally, potential applications of text analytics are discussed (including sentiment analysis or automatic generation of content).

R.-D. Kutsche and E. Zimányi (Eds.): eBISS 2019, LNBIP 390, pp. 25–39, 2020.
https://doi.org/10.1007/978-3-030-61627-4_2

2   Definition of Text Analytics

Text analytics, text data mining or text mining is the process of deriving information from textual sources. The input in the process is some text being a website, email, tweet or a document, and an output is in a format requested by a user being a chart, a list of features, an alert, etc. The goal is to transform text into meaningful data that may be used, e.g., in a decision process[1].

Typically, text analytics is used for several purposes, examples of which include:

– text summarization – trying to find the key content across a larger body of information or in a single document,
– document retrieval – retrieving documents referring to a concept or containing a specific phrase,
– sentiment analysis – identifying what is the nature of a commentary on an issue,
– event extraction – informing on events described within the text,
– explanation– finding what is the key issue driving a commentary,
– investigation – investigating what are the particular cases of a specific issue included and described within a text,
– classification – focusing on a subject or key content pieces the text talks about, enabling for grouping of documents.

To address these goals, various approaches were designed and implemented to identify the key concepts or emotional attitude of a text author to these concepts. The simplest form of text analytics involves extraction of keywords to create "bag of words" and developing a cloud of keywords. More sophisticated approaches include, e.g., named entities extraction, theme extraction, concept extraction, or sentiment analysis [8, 9].

Dealing with texts means addressing the data quality issue, e.g., spelling errors, grammar errors. When people refer to Named Entities being brands or product names, they sometimes write these names from capital letters, sometimes with a dash, sometimes with some spelling errors, etc. Other challenges include, e.g., free word order, homonyms, rich inflection, lack of data model behind the textual content. Another issue concerns a comparison between outcomes of an analysis. How to compare two different documents? How to deal with the multidimensionality (the more words or phrases, the more dimensions and the more complex is the analysis)?

These challenges impose certain requirements on each step in the text analysis process and are referred to while discussing how the text analysis should be carried out.

3   Sources of Textual Data

Text to be analysed comes from monitoring of diverse data sources and includes, e.g., HTML pages, RSS feeds, Facebook feeds, blogs, ... How to find these

---

[1] https://en.wikipedia.org/wiki/Text_mining.

sources? In many scenarios, sources that are monitored are the ones that have significant number of users or are known to provide insights into a specific domain. Such sources are either known to domain experts or are present in the first 10 links on Google search results. Finding a proper source of data is out of scope fo this paper, but some details may be found in [2].

Text is always more difficult to analyse than numerical data in databases, but still there are less and more difficult texts to analyse. For example, if text was automatically generated in the process of filling in a template with data from a database, we may use patterns based on a set of historical documents to extract the meaningful content. On the other hand, we may deal with a customer review written on a mobile phone while commuting that will be full or errors, inconsistencies or automatically-corrected words. Different types of texts usually need different approaches in the text processing process. Here, also the domain-specifics needs to be addressed - different domains have their own terminology and also phrasing may be greatly influenced.

The typical data sources being subject to the process of text analysis include:

- documents created automatically based on predefined templates and including data from structured data sources,
- documents (formal) created by humans following (or not) a certain template,
- documents published on the Web, e.g., press articles, product descriptions, legal documents, etc.,
- blog entries being articles published online,
- tweets, customer reviews, comments, etc.

It should be noted that the more user-related the content is, the more difficult it is to analyse. However, the more the content reflects the user opinion, the more valuable it is for a company or an entity.

4   Processing of Texts: The Pipeline

The text processing pipeline depends on the challenge to be solved. Sometimes, texts are not even initially pre-processed before an alert is created based on their content. However, a typical process of the data analysis consists of the following steps:

1. Data parsing (preprocessing).
2. Text segmentation.
3. Named Entity Extraction.
4. Data refinement.
5. Data description/structuring (if needed).
6. Text analytics/application of a chosen method of analysis.
7. Text visualisation.
8. Preparing data for usage in application scenarios, e.g., document search/retrieval.

The following sections address each of these steps showing potential techniques that may be applied. Some more details on the process of text processing may be found in, e.g., [5].

4.1 Step 1. Data Parsing

Parsing is a process of structuring the structured/un-structured content (to enable for further analyses) and concerns "reading" texts such as webblogs, RSS feeds, XML files, HTML files with the goal to find the part of text that we care about. Parsing is about identification of a block of text and making it available for further analyses[2]. However, parsing may be also understood as a more complex process consisting of the following steps:

1. Retrieving a document from a given source.
2. Using regular expression filters to pre-process the data.
3. Detecting paragraphs/removing HTML tags.
4. Tokenisation and detection of sentences.
5. Stopwords removal.
6. Identification of wordforms and morphology.

Regular Expressions ("regex" or "regexp") provide concise and flexible means for matching strings of text, such as particular characters, words, or patterns of characters. A regular expression is written in a formal language that can be interpreted by a regular expression processor[3].

Regular expressions descend from a fundamental concept in Computer Science called finite automata theory. A regular expression describes a pattern to match multiple input strings and therefore the simplest regular expression is a string of literal characters to match. A string matches a regular expression, if it contains the sub-string described by a regular expression. A regular expression can match a string in more than one place in a given text.

Examples of regular expressions encompass:

- (abc)* matches abc, abcabc, abcabcabc, . . .
- (abc){2, 3} matches abcabc or abcabcabc
- Section [0–9]+ enables to find all numbered sections
- Section [0–9]+\.[0–9]+ enables to find also all subsections, e.g., Sect. 4.3
- (a*)|([bcd]+) describes the following strings a AND aaa BUT ALSO bbb AND cc AND d
- (bc|de)g describes the following strings abcg AND adeg AND ...

where:

- (...) represents characters to be captured as a group,
- [0−9] indicates a number,
- * means zero or more repetitions,
- {m, n} stands for m to n repetitions,
- + means one or more repetitions,
- ? indicates an optional character.

---

[2] https://en.wikipedia.org/wiki/Parsing.
[3] https://en.wikipedia.org/wiki/Regular_expression.

Why do we need regular expressions? While looking at a website using a Web browser we see nicely formatted content with pictures, but from the processing perspective this content being an HTML/Java Script code mixes the interesting text with code snippets, formatting instructions, metadata, etc. Using regular expressions, we may extract only interesting phrases/elements matching the regular expressions. When do we need regular expressions? If the task is to find strings or phrases matching a specific pattern either to identify a specific phrase (name of a product) or to remove some part of a text, e.g., html marks. In case of dealing with HTML/XML some other solution may be of use, e.g., XPath.

xPath (XML Path Language) is an alternative to regular expressions based on the Document Object Model (DOM) being a cross-platform and language-independent application programming interface that treats an HTML, XHTML or XML document as a tree structure where in each node there is an object representing a part of the document[4]. XPath being a W3C standard and a query language for selecting nodes from an XML document, describes paths to elements in XML in a similar way an operating system describes paths to files and may be also used to compute values (e.g., strings, numbers, or Boolean values) from the content of an XML document. A path that begins with a/represents an absolute path, starting from the top of the document, e.g., /html/head/title. A path that does not begin with a/represents a path starting from the current element. Example: *head/title*. A path that begins with // can start from anywhere in the document, e.g., *//body/h1* selects every element h1 that is a child of an element body. For example, in Fig. 1 element title contains text "My title" that should be retrieved. XPath would define then a path *html/head/title* to identify the correct leaf and a function text() to get its content.

When to use regular expressions and when XPath? It is usually not a choice as they suit different needs, with XPath being used to retrieve text blocks from websites or XML files and regular expressions applied to further clean these documents or to check, e.g., if a document contains a specific phrase.

4.2   Step 2. Text Segmentation

After the textual content is extracted, one may focus on structuring the text to enable for further processing [13]. Text is just a sequence of characters and the two types of text segmentation involve:

- Low-level text segmentation (performed at the initial stages of text processing): **tokenisation and sentence splitting**.
- High-level text segmentation:

- Segmentation of linguistic groups such as **Named Entities** or Noun Phrases.
- Grouping sentences and paragraphs into **discourse topics**.

---

[4] https://en.wikipedia.org/wiki/XPath.

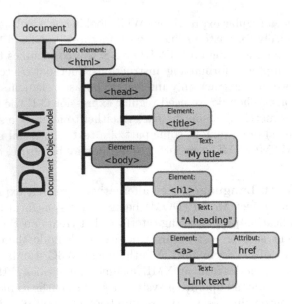

Fig. 1. An example of DOM. Source: https://en.wikipedia.org/wiki/Document_Obj ect_Model

Tokenisation is the process of segmenting text into linguistic units such as words, punctuation, numbers, alphanumerics, etc. Difficulty depends on the language as tokenisation in languages that are segmented is considered a relatively easy and uninteresting part of text processing (words delimited by blank spaces and punctuation), e.g., English. On the other hand in non-segmented languages, it is more challenging as no explicit boundaries between words are given, e.g., Chinese.

Sentence Splitting is the task of segmenting a text into sentences. The task may be perceived simple and as a general heuristic use punctuation marks such as . ? ! that usually signal a sentence boundary. The simplest algorithm that enables sentence splitting is known as 'period-space-capital letter'. It should be however noted that sometimes a period denotes a decimal point or is a part of an abbreviation. Therefore, lists of abbreviations, a lexicon of frequent sentence initial words and/or machine learning techniques shall be applied in case of more advanced scenarios in the group of Latin languages. In case of Chinese or Arabic different techniques need to be studied.

Removing Stopwords concerning removal of words that are the most common words in a language, e.g., "a", "and", "but", "how", "or", and "what". Stopwords are not content-bearing words and while analysis they introduce additional dimensions, so they are filtered out. Of course, in case of a full-text processing, stopwords should not be removed.

Part of Speech Tagging is the process of assigning a part-of-speech or lexical class marker, e.g., $[X]_{Noun}$ to each word in a corpus [7]. This may be due to the fact that nouns bring more value in terms of a topic of text, what may improve further processing.

Stemming is a reduction of as many related words and word forms as possible to a common canonical form – not necessarily the base form – which can then be used in the retrieval process. Term groups such as, e.g., CONNECT, CONNECTED, CONNECTING, CONNECTION, CONNECTIONS are conflated into a single term (by removal of the various suffixes -ED, -ING, -ION, -IONS to leave the single term CONNECT).

Lemmatisation is looking for a transformation to apply on a word to get its normalized form (identification of word endings: what word suffix should be removed and/or added to get the normalized form of a word). Lemmatisation is a process of grouping the inflected forms of a word together under a base form or of recovering the base form from an inflected form, e.g., grouping the inflected forms COME, COMES, COMING, CAME under the base form COME. The whole process is dictionary based. It takes a token (a word) and its part of speech information as an input. The output is a lemma of this word. The difference between stemming and lemmatisation is that stem might not be an actual word, but lemma is.

4.3 Step 3. Identification of Named Entities

Identification of Named Entities concerns a process aiming at finding proper names in texts and classifying these names into a set of predefined categories of interest, for example:

- entities: organizations, persons, locations,
- temporal expressions: time, date,
- quantities: monetary values, percentages, numbers.

There are two types of approaches to identification of named entities in text, often used together to enable achieving better insights:

- **Knowledge Engineering:** rule based, rules developed by experienced language engineers (time consuming as it requires manual work and rule coding).
- **Learning Systems:** use statistics or machine learning techniques to automatically learn the rules (requires large amounts of annotated training data).

More information on identification of Named Entities in text may be found in [10,12].

4.4 Step 4. Disambiguation

Disambiguation concerns selecting a sense for a word from a set of predefined possibilities. This sense usually comes from a dictionary (being a gazetter, a thesaurus or an ontology). Where is the challenge? When a person studying a text finds a word having one form, but several meanings, she extracts the sense based on a context. But from the perspective of an automated processing while having only a token, selecting a proper interpretation may be difficult when a word has several (sometimes contradictory meanings), e.g., title – a right of legal ownership, a document that is evidence of the legal ownership (closely related), a headline of a book or article.

Two phenomena related to this fact are called polysemy and homonymy. Polysemy concerns a situation where a single word form is associated with two or several related senses. In homonymy, a single word form is associated with two or several unrelated meanings [3].

There are three groups of approaches that enable dealing with the ambiguity:

– Knowledge-Based Disambiguation (use of external lexical resources such as dictionaries and thesauri),
– Supervised Disambiguation (based on a labelled training set),
– Unsupervised Disambiguation (based on unlabelled corpora, so we don't have proper labels for words in text).

Knowledge-based approaches attempt to disambiguate all open-class words in a text, e.g., "He put his suit over the back of the **chair**" using, e.g., information from dictionaries (definitions and examples for each meaning that enable to find similarity between definitions and current context), position in a semantic network (we may find that "table" is closer to "chair/furniture" than to "chair/person") or by using discourse properties (a word exhibits the same sense in a discourse/in a collocation).

Knowledge-based approaches are often used, e.g., for disambiguation of people names or geographical places. For example, in order to disambiguate if London refers to a capital of Great Britain or a city in Canada, one may apply the following heuristics:

– Relative importance of place (disambiguate based on the number of citizens).
– Comparison of a location to other places in the text (measuring distance between places). If a text is on sightseeing Canada, it probably refers to London in Canada.
– Context-based triggering, e.g., based on the name of a city mayor or event that is happening.

Supervised approaches concern learning to disambiguate words using annotated corpora. In this sense the disambiguation is viewed as a typical classification problem and therefore a training corpus is prepared and the rules are automatically discovered.

Unsupervised disambiguation means disambiguating without supporting tools such as dictionaries and thesauri and without a labelled training text.

Firstly, context vectors corresponding to all occurrences of a particular word are identified. Then, they are partitioned into regions of high density. In a next step the contexts of an ambiguous word are clustered to assign a sense to each such region (discriminate between these groups without actually labelling them).

One of the tools that may be used to support the process and that does not require advanced programming capabilities is Open Refine (formerly Google Refine)[5] that enables working with messy data: cleaning it, transforming it from one format into another and extending it with external data.

4.5   Step 5. Describing the Text

After processing, the document needs to be represented for the needs of using it in the future application scenarios. This includes creating the following representations:

- "Bag of words": common representation of texts in which all words from a document are represented, e.g., in a vector containing them. While representing a document as a "bag of words" stemming/lemmatisation are applied and stop words are eliminated to decrease the number of dimensions. Each word in a bag or words is presented only once, however frequency is often associated.
- Features of text: identifying title, keywords, date information, source information, named entities, what relates to creating metadata for the text concerning major features of an analysed document.
- Indexes: creating indexes of documents, e.g., nouns or phrases explaining what content the text described, coming directly from text or being derived from an ontological description of the text.

While representing a corpus (collection of documents) usually the indexing approach is applied as it enables for an efficient processing. A "reverse index" provides a way of keeping track of a list of all documents that contain a specific feature and for every possible feature, e.g., for every word/phrase a list of documents containing that word/phrase is identified.

When representing a document within the corpus to describe its importance from the perspective of a specific index, corpus-wide term frequency metrics are of use. Popular example here is **TF*IDF**. The concept behind the measure is that preferred indexes of text are terms frequent in a given document and rare in the whole collection. TF*IDF is a product of two statistics:

- TF: term frequency (the number of occurrences of a word in a document $n_{i,j}$ divided by the number of occurrences of all words in a document $\sum_k n_{k,j}$),
- IDF: inverse document frequency ($|D|$ is a number of documents in a corpus (collection of documents) and $|d : t_i \in d|$ – a number of documents with a given word (appearing at least once).

Figure 2 presents formulas for calculation of TF and IDF.

$$TF_{i,j} = \frac{n_{i,j}}{\sum_k n_{k,j}} \quad IDF_i = \log \frac{|D|}{|\{d : t_i \in d\}|}$$

Fig. 2. TF*IDF formulas.

Documents in a corpus are often represented as vectors or pairs of words with these words' frequency. Such vectors are analysed in a vector space, where terms are axes and docs are put somewhere in this space depending on the frequency of the terms in each of these documents (please see Fig. 3). Even, with a proper initial pre-processing this may lead to over 20.000 dimensions. Such representation of documents simplifies the problem of finding similar documents (in terms of content, not length) using the cosine measure. The idea behind the cosine measure is to measure the distance between vectors, e.g., D_1 and D_2 captured by the cosine of the angle θ between them (please note this is similarity, not a distance).

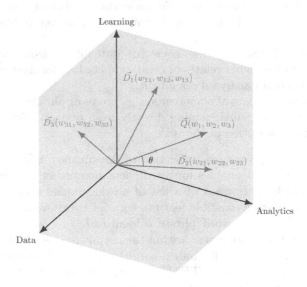

Fig. 3. Documents representation in a form of vectors. Source: [1]

The following example[6] explains the idea behind comparison of two documents. Having two documents (each represented by one line below):

Julie loves me more than Linda loves me.
Jane likes me more than Julie loves me.

---

[6] https://medium.com/@sumn2u/cosine-similarity-between-two-sentences-8f6630b0ebb7.

in the first step a list of words used in both documents is created: **me Julie Jane loves likes Linda than more**. Then, two vectors with frequencies of occurrence of the words in the texts are developed:

a: [2, 1, 0, 2, 0, 1, 1, 1]
b: [2, 1, 1, 1, 1, 0, 1, 1]

In the end, the cosine measure is applied using the formula:

$$cosine = \frac{\sum_{i=1}^{n}(a_i \cdot b_i)}{(\sqrt{\sum_{i=1}^{n} a^2} \cdot \sqrt{\sum_{i=1}^{n} b^2})}$$

which in the discussed case is 0,822.

4.6   Step 6: Analytics: Topic Tagging

The goal of the topic extraction is to tag names of people, places or organizations in any type of content, in order to make them more findable and linkable to other contents. Unfortunately, the topic tagging highly depends on the type of content that is analysed. Example heuristics that may be applied to this tasks involve:

- Counting a number of occurrences of a given named entity in the document.
- Checking the document features: is the entity name in the title of the document, how often it appears in the text, are there any abbreviations that are used, etc.
- Identifying hashtags, e.g., in tweets.
- Deriving topic based on classification rules that were previously trained.
- Identifying the frequently used named entities with their polarity metrics.

5   Application Scenarios

The text analysis is always implemented bearing in mind scenarios that are to be implemented. Two most frequent use cases involve sentiment analysis and search.

5.1   Sentiment Analysis

Sentiment is a view or attitude towards and object, based on emotion instead of a reason. The attitude may be positive, negative or neutral, or when following another classification it may refer to emotions, e.g., anger, sadness, happiness. Sentiment Analysis (opinion mining) is application of NLP (natural language processing) to extraction or classification of sentiment from typically unstructured text, e.g., reviews, tweets [6].

While analysing documents with the goal of sentiment study, one identifies:

- **opinion holder:** a person or organization that holds a specific opinion on a particular object,

- **object:** on which an opinion is expressed,
- **opinion:** a view, attitude, or appraisal on an object from an opinion holder.

The sentiment analysis may be performed studying a document with goals and from different perspectives, e.g.,:

- On the **document (or review) level** the task is to classify the documents. The assumption behind is that each document (or review) focuses on a single object (not true in many discussion posts) and contains opinion from a single opinion holder.
- On the **sentence level** subjective sentences may be identified to enable quick reaction or detailed sentiment may be studied.
- On the **feature level** the task may be to identify and extract object features that have been commented on by an opinion holder (e.g., a reviewer), to determine whether the opinions on the features are positive, negative or neutral and as the last step to group feature synonyms producing a feature-based opinion summary of multiple reviews.

As a result of such an analysis based on a document we may derive, e.g.:

- sentiment towards a certain feature of a product,
- sentiment expressed in the whole document,
- sentiment expressed in a collection of documents,
- comparison of views on a product,
- comparison of our brand with our competitor.

Sentiment analysis is more difficult than topical classification, with which bag of words performs well. This mainly due to the fact that it must consider other features due to subtlety of sentiment expression, irony and expression of sentiment using neutral words and is greatly domain dependent (words/phrases can mean different things in different contexts and domains).

5.2   Search and Retrieval

Search and retrieval of documents is about answering the following questions:

- Which documents have this word or this phrase?
- Which documents concern this topic or this entity?

The collection of texts to be processed needs to be initially processed to enable for searching. The techniques originate from the field of library science and usually concern as their foundation creation of the previously described inverted index.

However, when searching not only the initial preparation of documents to be found is important, but also future assessment of results achieved. The metrics that may be used to assess the quality of search include:

- Relevance: is this document what I wanted? Is a document a relevant answer to my query?

- Precision: what % of documents in the result are relevant as a response to a query (are referring to the topic mentioned in a query)?
- Recall: of all the relevant documents in the corpus, what % were returned to me?
- "Authoritativeness": how authoritative a source is? How many other sources refer to the source?
- Recency of a document: when was the document created (new documents are more relevant than old ones)?
- Popularity: how often the document has been retrieved by other users?

Please note that authoritativeness, recency or popularity of documents may be also easily combined while preparing a response to a query posed by a user (together with TF*IDF metric).

6   Case Study

Nowadays, a frequent application scenario for text analysis is brand management. This concerns monitoring Internet sources to learn inter alia:

- Are people mentioning the brand and products?
- What do people say? Is it positive or negative?
- And how about the products of the competitors?

Table 1 presents how a process of brand management could look like. Specific tasks are identified and exemplary methods to be applied are mentioned.

Table 1. Text analytics for the needs of brand management.

Task/Goal	Method to be applied
1. Monitor social networks, review sites for mentions of products	Parse the data feeds to get the content. Find the product names (using, e.g., regular expressions or Named Entity Recognition)
2. Collect the reviews	Extract the relevant text. Convert the text into a suitable representation, e.g., indexes
3. Sort the reviews by product	Classification (Topic Tagging)
4. Are they bad or good?	Classification (Sentiment Analysis)
5. Marketing department reads selected reviews in full to get an insight	Search/Information Retrieval

7 Summary

Text analysis is a complex process which steps heavily depend on the analytical task that is to be achieved. Before applying any of the previously described methods, it is important to identify a problem that is to be solved and align the process to the needs of addressing the problem. The following step will concern finding the right structure for the unstructured data and selecting the proper analysis method. It should be explicitly written that many known methods such as k-NN and k-Means may work well in some of the scenarios.

Summarising, the paper presented the typical process of text analysis. It started from the presentation of challenges within text analysis, especially emerging from the data quality. Then, the paper discusses use of regular expressions and XPath in the text parsing. Following, key tasks in text analysis are presented. In the concluding part of the paper application scenarios are presented.

References

1. Chapter 11 - Information retrieval: Concepts, models, and systems. In: Gudivada, V.N., Rao, C. (eds.) Computational Analysis and Understanding of Natural Languages: Principles, Methods and Applications. Handbook of Statistics, vol. 38, pp. 331–401. Elsevier (2018)
2. Cooper, D., Schindler, P.: Business Research Methods. McGraw-Hill, New York (2016)
3. Falkum, I., Vicente, A.: Polysemy: current perspectives and approaches. Lingua **157**, 02 (2015)
4. Fang, X., Zhan, J.: Sentiment analysis using product review data. J. Big Data **2**(1), 1–14 (2015). https://doi.org/10.1186/s40537-015-0015-2
5. Gudivada, V.N.: Chapter 12 - Natural language core tasks and applications. In: Gudivada, V.N., Rao, C. (eds.) Computational Analysis and Understanding of Natural Languages: Principles, Methods and Applications. Handbook of Statistics, vol. 38, pp. 403–428. Elsevier (2018)
6. Hussein, D.M.E.-D.M.: A survey on sentiment analysis challenges. J. King Saud Univ. Eng. Sci. **30**(4), 330–338 (2018)
7. Jurafsky, D., Martin, J.: Speech and Language Processing: An Introduction to Natural Language Processing, Computational Linguistics, and Speech Recognition, vol. 2 (2008)
8. Kane, D.: Data science - PART XI - text analytics. https://www.datasciencecentral.com/profiles/blogs/an-introduction-to-text-analytics. Accessed 01 Oct 2019
9. Kotu, V., Deshpande, B.: Chapter 9 - Text mining. In: Kotu, V., Deshpande, B. (eds.) Data Science, 2nd edn., pp. 281–305. Morgan Kaufmann, Burlington (2019)
10. Nadeau, D., Sekine, S.: A survey of named entity recognition and classification. Linguisticae Investigationes **30**(1), 3–26 (2007)
11. Sonntag, D.: Assessing the quality of natural language text data. In: INFORMATIK 2004 - Informatik verbindet, Band 1, Beiträge der 34. Jahrestagung der Gesellschaft für Informatik e.V. (GI), Ulm, 20–24 September 2004, pp. 259–263 (2004)

12. Yadav, V., Bethard, S.: A survey on recent advances in named entity recognition from deep learning models. In: Proceedings of the 27th International Conference on Computational Linguistics, Santa Fe, New Mexico, USA, pp. 2145–2158. Association for Computational Linguistics, August 2018
13. Yordanov, V.: Introduction to natural language processing for text. https://towardsdatascience.com/introduction-to-natural-language-processing-for-text-df845750fb63. Accessed 01 Oct 2019

Automated Machine Learning:
Techniques and Frameworks

Radwa Elshawi$^{(\boxtimes)}$ and Sherif Sakr

Data Systems Group, University of Tartu, Tartu, Estonia
{radwa.elshawi,sherif.sakr}@ut.ee

Abstract. Nowadays, machine learning techniques and algorithms are employed in almost every application domain (e.g., financial applications, advertising, recommendation systems, user behavior analytics). In practice, they are playing a crucial role in harnessing the power of massive amounts of data which we are currently producing every day in our digital world. In general, the process of building a high-quality machine learning model is an iterative, complex and time-consuming process that involves trying different algorithms and techniques in addition to having a good experience with effectively tuning their hyper-parameters. In particular, conducting this process efficiently requires solid knowledge and experience with the various techniques that can be employed. With the continuous and vast increase of the amount of data in our digital world, it has been acknowledged that the number of knowledgeable data scientists can not scale to address these challenges. Thus, there was a crucial need for automating the process of building good machine learning models (AutoML). In the last few years, several techniques and frameworks have been introduced to tackle the challenge of automating the machine learning process. The main aim of these techniques is to reduce the role of humans in the loop and fill the gap for non-expert machine learning users by playing the role of the domain expert. In this chapter, we present an overview of the state-of-the-art efforts in tackling the challenges of machine learning automation. We provide a comprehensive coverage for the various tools and frameworks that have been introduced in this domain. In addition, we discuss some of the research directions and open challenges that need to be addressed in order to achieve the vision and goals of the AutoML process.

1 Introduction

Due to the increasing success of machine learning techniques in several application domains, they have been attracting a lot of attention from the research and business communities. In general, the effectiveness of machine learning techniques mainly rests on the availability of massive datasets. Recently, we have been witnessing a continuous exponential growth in the size of data produced by various kinds of systems, devices and data sources. It has been reported that there are 2.5 quintillion bytes of data is being created every day where 90% of

© Springer Nature Switzerland AG 2020
R.-D. Kutsche and E. Zimányi (Eds.): eBISS 2019, LNBIP 390, pp. 40–69, 2020.
https://doi.org/10.1007/978-3-030-61627-4_3

stored data in the world, has been generated in the past two years only[1]. On the one hand, the more data that is available, the richer and the more robust the insights and the results that machine learning techniques can produce. Thus, in the Big Data Era, we are witnessing many leaps achieved by machine and deep learning techniques in a wide range of fields [1,2]. On the other hand, this situation is raising a potential *data science crisis*, similar to the software crisis [3], due to the crucial need of having an increasing number of data scientists with strong knowledge and good experience so that they are able to keep up with harnessing the power of the massive amounts of data which are produced daily. In particular, it has been acknowledged that *data scientists can not scale*[2] and it is almost impossible to balance between the number of qualified data scientists and the required effort to manually analyze the increasingly growing sizes of available data. Thus, we are witnessing a growing focus and interest to support automating the process of building machine learning pipelines where the presence of a human in the loop can be dramatically reduced, or preferably eliminated.

In general, the process of building a high-quality machine learning model is an iterative, complex and time-consuming process that involves a number of steps. In particular, a data scientist is commonly *challenged* with a large number of choices where informed decisions need to be taken. For example, the data scientist needs to select among a wide range of possible algorithms including classification or regression techniques (e.g. Support Vector Machines, Neural Networks, Bayesian Models, Decision Trees, etc.) in addition to tuning numerous hyper-parameters of the selected algorithm. In addition, the performance of the model can also be judged by various metrics (e.g., accuracy, sensitivity, specificity, F1-score). Naturally, the decisions of the data scientist in each of these steps affect the performance and the quality of the developed model [4–6]. For instance, in `yeast dataset`[3], different parameter configurations of a Random Forest classifier result in different range of accuracy values, around 5%[4]. Also, using different classifier learning algorithms leads to widely different performance values, around 20%, for the fitted models on the same dataset. Although making such decisions require solid knowledge and expertise, in practice, increasingly, users of machine learning tools are often non-experts who require *off-the-shelf* solutions. Therefore, there has been a growing interest to *automate* and *democratize* the steps of building the machine learning pipelines.

In the last years, several techniques and frameworks have been introduced to tackle the challenge of automating the process of Combined Algorithm Selection and Hyper-parameter tuning (CASH) in the machine learning domain. These techniques have commonly formulated the problem as an optimization problem that can be solved by a wide range of techniques [7–9]. In general, the *CASH* problem is described as follows:

---

[1] Forbes: How Much Data Do We Create Every Day? May 21, 2018.
[2] https://hbr.org/2015/05/data-scientists-dont-scale.
[3] https://www.openml.org/d/40597.
[4] https://www.openml.org/t/2073.

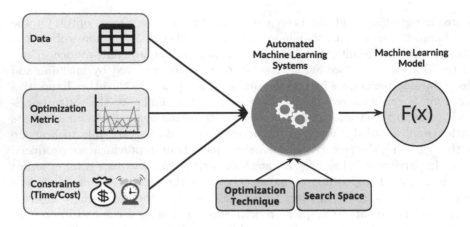

Fig. 1. The general workflow of the AutoML process.

Given a set of machine learning algorithms $\mathbf{A} = \{A^{(1)}, A^2, ...\}$, and a dataset D divided into disjoint training D_{train}, and validation $D_{validation}$ sets. The goal is to find an algorithm $A^{(i)^*}$ where $A^{(i)} \in \mathbf{A}$ and $A^{(i)^*}$ is a tuned version of $A^{(i)}$ that achieves the highest generalization performance by training $A^{(i)}$ on D_{train}, and evaluating it on $D_{validation}$. In particular, the goal of any CASH optimization technique is defined as:

$$A^{(i)^*} \in \underset{A \in \mathbf{A}}{argmin}\ L(A^{(i)}, D_{train}, D_{validation})$$

where $L(A^{(i)}, D_{train}, D_{validation})$ is the loss function (e.g.: error rate, false positives, etc.). In practice, one constraint for CASH optimization techniques is the *time budget*. In particular, the aim of the optimization algorithm is to select and tune a machine learning algorithm that can achieve (near)-optimal performance in terms of the user-defined evaluation metric (e.g., accuracy, sensitivity, specificity, F1-score) within the user-defined *time budget* for the search process (Fig. 1).

In this chapter, we present an overview of the state-of-the-art efforts for the techniques and framework in the automated machine learning domain. The remainder of this chapter is organized as follows. Section 2 covers the various techniques and frameworks that have been introduced to tackle the challenge of the automated machine learning process while Sect. 3 covers the automated deep learning process. We discuss some of the research directions and open challenges that need to be addressed in order to achieve the vision and goals of the AutoML process in Sect. 4 before we finally conclude the chapter in Sect. 5.

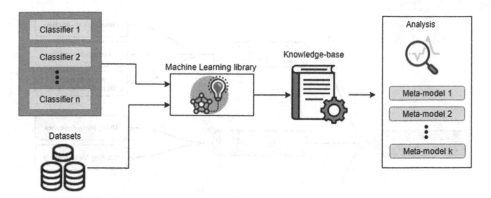

Fig. 2. An overview of meta-learning process.

2 Automated Machine Learning

In general, meta-learning can be described as the process of learning from previous experience gained during applying various learning algorithms on different kinds of data, and hence reducing the needed time to learn new tasks [10]. In the context of machine learning, several *meta learning*-techniques have been introduced as an effective mechanism to tackle the challenge of warm start for optimization algorithms. Figure 2 illustrates an overview of the meta-learning process. These techniques can generally be categorized into three broad groups [11]: *learning based on task properties, learning from previous model evaluations* and *learning from already pretrained models* (Fig. 3).

One group of meta-learning techniques has been based on learning from task properties using the *meta-features* that characterize a particular dataset [9]. Generally speaking, each prior task is characterized by a feature vector, of k features, $m(t_j)$. Simply, information from a prior task t_j can be transferred to a new task t_{new} based on their similarity, where this similarity between t_{new} and t_j can be calculated based on the distance between their corresponding feature vectors. In addition, a meta learner L can be trained on the feature vectors of prior tasks along with their evaluations $\mathbf{P}$ to predict the performance of configurations θ_i on t_{new}.

Some of the commonly used meta features for describing datasets are simple meta features including number of instances, number of features, statistical features (e.g., skewness, kurtosis, correlation, co-variance, minimum, maximum, average), landmark features (e.g., performance of some landmark learning algorithms on a sample of the dataset), and information theoretic features (e.g., the entropy of class labels) [11]. In practice, the selection of the best set of meta features to be used is highly dependent on the application [12]. When computing the similarity between two tasks represented as two feature vectors of meta data, it is important to normalize these vectors or apply dimensionality reduction techniques such as principle component analysis [12,13]. Another way to

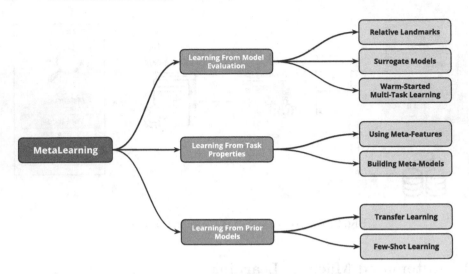

Fig. 3. A taxonomy of meta-learning techniques.

extract meta-features is to learn a joint distribution representation for a set of tasks.

Another meta-learning approach is to learn from prior tasks properties is through building *meta-models*. In this process, the aim is to build a meta model L that learns complex relationships between meta features of prior tasks t_j. For a new task t_{new}, given the meta features for task t_{new}, model L is used to recommend the best configurations. There exists a rich literature on using meta models for model configuration recommendations [14–18]. Meta models can also be used to rank a particular set of configurations by using the $K-$nearest neighbour model on the meta features of prior tasks and predicting the top k tasks that are similar to new task t_{new} and then ranking the best set of configurations of these similar tasks [19,20]. Moreover, they can also be used to predict the performance of new task based on a particular configuration [21,22]. This gives an indication about how good or bad this configuration can be, and whether it is worth evaluating it on a particular new task.

Another group of meta-learning techniques are based on *learning from previous model evaluation*. In this context, the problem is formally defined as follows.

Given a set of machine learning tasks $t_j \in T$, their corresponding learned models along their hyper-parameters $\theta \in \Theta$ and $P_{i,j} = P(\theta_i, t_j)$, the problem is to learn a meta-learner L that is trained on meta-data $\mathbf{P} \cup \mathbf{P}_{new}$ to predict recommended configuration Θ_{new}^* for a new task t_{new}, where T is the set of all prior machine learning tasks. Θ is the configuration space (hyper-parameter setting, pipeline components, network architecture, and network hyper-parameter), Θ_{new} is the configuration space for a new machine learning task t_{new}, $\mathbf{P}$ is the set of all prior evaluations $P_{i,j}$ of configuration θ_i on a prior task t_j, and $\mathbf{P}_{new}$ is a set of evaluations $P_{i,new}$ for a new task t_{new}.

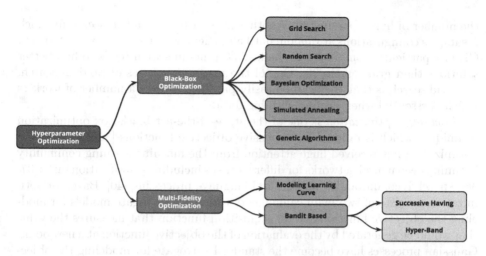

Fig. 4. A taxonomy for the hyper-parameter optimization techniques.

Learning from prior models can be done using *Transfer learning* [23], which is the process of utilization of pretrained models on prior tasks t_j to be adapted on a new task t_{new}, where tasks t_j and t_{new} are similar. Transfer learning has received lots of attention especially in the area of neural network. In particular, neural network architecture and neural network parameters are trained on prior task t_j that can be used as an initialization for model adaptation on a new task t_{new}. Then, the model can be fine-tuned [24–26]. It has been shown that neural networks trained on big image datasets such as ImageNet [17] can be transferred as well to new tasks [27,28]. Transfer learning usually works well when the new task to be learned is similar to the prior tasks, otherwise transfer learning may lead to unsatisfactory results [29]. In addition, prior models can be used in *Few-Shot Learning* where a model is required to be trained using a few training instances given the prior experience gained from already trained models on similar tasks.

2.1 Hyper-parameter Optimization

In general, several hyper-parameter optimization techniques have been based and borrowed ideas from the domains of statistical model selection and traditional optimization techniques [30–32]. In principle, the automated hyper-parameter tuning techniques can be classified into two main categories: *black-box optimization techniques* and *multi-fidelity optimization techniques* (Fig. 4).

Black-Box Optimization. *Grid search* is a simple basic solution for the hyper-parameter optimization [33] in which all combinations of hyper-parameters are evaluated. Thus, grid search is computationally expensive, infeasible and suffers from the *curse of dimensionality* as the number of trails grows exponentially with

the number of hyper-parameters. Another alternative is *random search* in which it samples configurations at random until a particular budget B is exhausted [34]. Given a particular computational budget B, random search tends to find better solutions than grid search [33]. One of the main advantages of random search, and grid search is that they can be easily parallelized over a number of workers which is essential when dealing with big data.

Bayesian Optimization is one of the state-of-the-art black-box optimization techniques which is tailored for expensive objective functions [35, 36]. Bayesian optimization has received huge attention from the machine learning community in tuning deep neural networks for different tasks including classification tasks [37, 38], speech recognition [39] and natural language processing [40]. Bayesian optimization consists of two main components which are surrogate models for modeling the objective function and an acquisition function that measures the value that would be generated by the evaluation of the objective function at a new point. Gaussian processes have become the standard surrogate for modeling the objective function in Bayesian optimization [38, 41]. One of the main limitations of the Gaussian processes is the cubic complexity to the number of data points which limits their parallelization capability. Another limitation is the poor scalability when using the standard kernels. Random forests [42] are another choice for modeling the objective function in Bayesian optimization. First, the algorithm starts with growing B regression trees, each of which is built using n randomly selected data points with replacement from training data of size n. For each tree, a split node is chosen from d algorithm parameters. The minimum number of points are considered for further split are set to 10 and the number of trees B to grow is set be 10 to maintain low computational overhead. Then, the random forest predicted mean and variance for each new configuration is computed. The random forests' complexity of the fitting and predicting variances are $O(n \log n)$ and $O(\log n)$ respectively which is much better compared to the Gaussian process. Random forests are used by the Sequential Model-based Algorithm Configuration (SMAC) library [43]. In general Tree-structured Parzen Estimator (TPE) [44] does not define a predictive distribution over the objective function but it creates two density functions that act as generative models for all domain variables. Given a percentile α, the observations are partitioned into two sets of observations (good observations and bad observations) where simple Parzen windows are used to model the two sets. The ratio between the two density functions reflects the expected improvement in the acquisition function and is used to recommend new configurations for hyper-parameters. Tree-Structured Parzen estimator (TPE) has shown great performance for hyper-parameter optimization tasks [44–48].

Simulated Annealing is a hyper-parameter optimization approach which is inspired by the metallurgy technique of heating and controlled cooling of materials [49]. This optimization technique goes through a number of steps. First, it randomly chooses a single value (current state) to be applied to all hyper-parameters and then evaluates model performance based on it. Second, it randomly updates the value of one of the hyper-parameters by picking a value from the immediate neighborhood to get neighboring state. Third, it evaluates the model performance

based on the neighboring state. Forth, it compares the performance obtained from the current and neighbouring states. Then, the user chooses to reject or accept the neighbouring state as a current state based on some criteria.

Genetic Algorithms (GA) are inspired by the process of natural selection [50]. The main idea of genetic-based optimization techniques is simply applying multiple genetic operations to a population of configurations. For example, the *crossover* operation simply takes two parent *chromosomes* (configurations) and combines their genetic information to generate new *offspring*. More specifically, the two parents configurations are cut at the same crossover point. Then, the sub-parts to the right of that point are swapped between the two parents chromosomes. This contributes to two new *offspring* (child configuration). Mutation randomly chooses a chromosome and mutates one or more of its parameters that results in a totally new chromosome.

Multi-fidelity Optimization. Multi-fidelity optimization is an optimization technique which focuses on decreasing the evaluation cost by combining a large number of cheap low-fidelity evaluations and a small number of expensive high-fidelity evaluation [51]. In practice, such an optimization technique is essential when dealing with big datasets as training one hyper-parameter may take days. More specifically, in multi-fidelity optimization, we can evaluate samples in different levels. For example, we may have two evaluation functions: *high-fidelity* evaluation and *low-fidelity* evaluation. The high-fidelity evaluation outputs precise evaluation from the whole dataset. On the other hand, the low-fidelity evaluation is a cheaper evaluation from a subset of the dataset. The idea behind the multi-fidelity evaluation is to use many low-fidelity evaluation to reduce the total evaluation cost. Although the low fidelity optimization results in cheaper evaluation cost that may suffer from optimization performance, but the speedup achieved is more significant than the approximation error.

Modeling learning curves is an optimization technique that models learning curves during hyper-parameter optimization and decides whether to allocate more resources or to stop the training procedure for a particular configuration. For example, a curve may model the performance of a particular hyper-parameter on an increasing subset of the dataset. Learning curve extrapolation is used in predicting early termination for a particular configuration [36]; the learning process is terminated if the performance of the predicted configuration is less than the performance of the best model trained so far in the optimization process. Combining early predictive termination criterion with Bayesian optimization leads to more reduction in the model error rate than the vanilla Bayesian black-box optimization. In addition, such a technique resulted in speeding-up the optimization by a factor of 2 and achieved the state-of-the-art neural network on CIFAR-10 dataset [52].

Bandit-based algorithms have shown to be powerful in tackling deep learning optimization challenges. In the following, we consider two strategies of the bandit-based techniques which are the *Successive halving* and *HyperBand*. *Successive halving* is a bandit-based powerful multi-fidelity technique in which

given a budget B, first, all the configurations are evaluated. Next, they are ranked based on their performance. Then, half of these configurations that performed worse than the others are removed. Finally, the budget of the previous steps is doubled and repeated until only one algorithm remains. It is shown that the successive halving outperforms the uniform budget al.location technique in terms of the computation time, and the number of iterations required [53]. On the other hand, successive halving suffer from the following problem. Given a time budget B, the user has to choose, in advance, whether to consume the larger portion of the budget exploring a large number of configurations while spending a small portion of the time budget on tuning each of them or to consume the large portion of the budget on exploring few configurations while spending the larger portion of the budget on tuning them.

HyperBand is another bandit-based powerful multi-fidelity hedging technique that optimizes the search space when selecting from randomly sampled configurations [54]. More specifically, partition a given budget B into combinations of number of configurations and budget assigned to each configuration. Then, call successive halving technique on each random sample configuration. Hyper-Band shows great success with deep neural networks and performs better than random search and Bayesian optimization.

2.2   AutoML Tools and Frameworks

In this section, we provide a comprehensive overview of several tools and frameworks that have been implemented to automate the process of combined algorithm selection and hyper-parameter optimization process. In general, these tools and frameworks can be classified into two main categories: *centralized* and *distributed*.

Centralized Frameworks. Several tools have been implemented on top of widely used *centralized* machine learning packages which are designed to run in a *single* node (machine). In general, these tools are suitable for handling small and medium sized datasets. For example, *Auto-Weka*[5] is considered as the first and pioneer machine learning automation framework [7]. It was implemented in Java on top of `Weka`[6], a popular machine learning library that has a wide range of machine learning algorithms. `Auto-Weka` applies Bayesian optimization using Sequential Model-based Algorithm Configuration (`SMAC`) [43] and tree-structured parzen estimator (TPE) for both algorithm selection and hyper-parameter optimization (Auto-Weka uses SMAC as its default optimization algorithm but the user can configure the tool to use TPE). In particular, SMAC tries to draw the relation between algorithm performance and a given set of hyper-parameters by estimating the predictive mean and variance of their performance along the trees of a random forest model. The main advantage of using SMAC is its robustness by having the ability to discard low performance parameter configurations

---

[5] https://www.cs.ubc.ca/labs/beta/Projects/autoweka/.
[6] https://www.cs.waikato.ac.nz/ml/weka/.

quickly after the evaluation on a low number of dataset folds. SMAC shows better performance on experimental results compared to TPE [43].

$Auto-MEKA_{GGP}$ [55] focuses on the AutoML task for multi-label classification problem [56] that aims to learn models from data capable of representing the relationships between input attributes and a set of class labels, where each instance may belong to more than one class. Multi-label classification has lots of applications especially in medical diagnosis in which a patient may be diagnosed with more than one disease. $Auto-MEKA_{GGP}$ is a grammar-based genetic programming framework that can handle complex multi-label classification search space and simply explores the hierarchical structure of the problem. $Auto-MEKA_{GGP}$ takes as input both of the dataset and a grammar describing the hierarchical search space of the hyper-parameters and the learning algorithms from MEKA[7] framework [57]. $Auto-MEKA_{GGP}$ starts by creating an initial set of trees representing the multi-label classification algorithms by randomly choosing valid rules from the grammar, followed by the generation of derivation trees. Next, map each derivation tree to a specific multi-label classification algorithm. The initial trees are evaluated on the input dataset by running the learning algorithm, they represent, using MEKA framework. The quality of the individuals are assessed using different measures such as fitness function. If a stopping condition is satisfied (e.g. a quality criteria), a set of individuals (trees) are selected in a tournament selection. Crossover and mutation are applied in a way that respects the grammar constraints on the selected individuals to create a new population. At the end of the evolution, the best set of individuals representing the well performing set of multi-label tuned classifiers are returned.

$Auto-Sklearn$[8] has been implemented on top of Scikit-Learn[9], a popular Python machine learning package [8]. Auto-Sklearn introduced the idea of meta-learning in the initialization of combined algorithm selection and hyper-parameter tuning. It used SMAC as a Bayesian optimization technique too. In addition, ensemble methods were used to improve the performance of output models. Both meta-learning and ensemble methods improved the performance of *vanilla* SMAC optimization. *hyperopt-Sklearn* [58] is another AutoML framework which is based on Scikit-learn machine learning library. Hyperopt-Sklearn uses Hyperopt [59] to define the search space over the possible Scikit-Learn main components including the learning and preprocessing algorithms. Hyperpot supports different optimization techniques including random search, and different Bayesian optimizations for exploring the search spaces which are characterized by different types of variables including categorical, ordinal and continuous.

$TPOT$[10] framework represents another type of solution that has been implemented on top of Scikit-Learn [60]. It is based on genetic programming by exploring many different possible pipelines of feature engineering and learning algorithms. Then, it finds the best one out of them. *Recipe* [61] follows the same optimization procedure as TPOT using genetic programming, which in turn

---

[7] http://waikato.github.io/meka/.
[8] https://github.com/automl/auto-sklearn.
[9] https://scikit-learn.org/.
[10] https://automl.info/tpot/.

exploits the advantages of a global search. However, it considers the unconstrained search problem in TPOT, where resources can be spent into generating and evaluating invalid solutions by adding a grammar that avoids the generation of invalid pipelines, and can speed up optimization process. Second, it works with a bigger search space of different model configurations than *Auto-SkLearn* and TPOT.

ML-Plan[11] has been proposed to tackle the composability challenge on building machine learning pipelines [62]. In particular, it integrates a super-set of both Weka and Scikit-Learn algorithms to construct a full pipeline. ML-Plan tackles the challenge of the search problem for finding optimal machine learning pipeline using a hierarchical task network algorithm where the search space is modeled as a large tree graph where each leaf node is considered as a goal node of a full pipeline. The graph traversal starts from the root node to one of the leaves by selecting some random paths. The quality of a certain node in this graph is measured after making n such random complete traversals and taking the minimum as an estimate for the best possible solution that can be found. The initial results of this approach has shown that the composable pipelines over Weka and Scikit-Learn do not significantly outperform the outcomes from Auto-Weka and Auto-Sklearn frameworks because it has to deal with larger search space.

SmartML[12] has been introduced as the first R package for automated model building for classification tasks [9]. In the algorithm selection phase, SmartML uses a meta-learning approach where the meta-features of the input dataset is extracted and compared with the meta-features of the datasets that are stored in the framework's knowledge base, populated from the results of the previous runs. The similarity search process is used to identify the similar datasets in the knowledge base, using a nearest neighbor approach, where the retrieved results are used to identify the best performing algorithms on those similar datasets in order to nominate the candidate algorithms for the dataset at hand. The hyper-parameter tuning of SmartML is based on SMAC Bayesian Optimisation [43]. SmartML maintains the results of the new runs to continuously enrich its knowledge base with the aim of further improving the accuracy of the similarity search and thus the performance and robustness for future runs.

Autostacker [63] is an AutoML framework that uses an evolutionary algorithm with hierarchical stacking for efficient hyper-parameters search. Autostacker is able to find pipelines, consisting of preprocessing, feature engineering and machine learning algorithms with the best set of hyper-parameters, rather than finding a single machine learning model with the best set of hyper-parameters. Autostacker generates cascaded architectures that allow the components of a pipeline to "correct mistakes made by each other" and hence improves the overall performance of the pipeline. Autostacker simply starts by selecting a set of pipelines randomly. Those pipelines are fed into an evolutionary algorithm that generates the set of winning pipelines.

AlphaD3M [64] has been introduced as an AutoML framework that uses meta reinforcement learning to find the most promising pipelines. AlphaD3M finds

---

[11] https://github.com/fmohr/ML-Plan.
[12] https://github.com/DataSystemsGroupUT/SmartML.

patterns in the components of the pipelines using recurrent neural networks, specifically long short term memory (LSTM) and Monte-Carlo tree search in an iterative process which is computationally efficient in large search space. In particular, for a given machine learning task over a certain dataset, the network predicts the action's probabilities which lead to sequences that describe the whole pipeline. The predictions of the LSTM neural network are used by Monte-Carlo tree search by running multiple simulations to find the best pipeline sequence.

$OBOE$[13] is an AutoML framework for time constrained model selection and hyper-parameter tuning [65]. OBOE finds the most promising machine learning model along with the best set of hyper-parameters using collaborative filtering. OBOE starts by constructing an *error matrix* for some base set of machine learning algorithms, where each row represents a dataset and each column represents a machine learning algorithm. Each cell in the matrix represents the performance of a particular machine learning model along with its hyper-parameters on a specific dataset. In addition, OBOE keeps track of the running time of each model on a particular dataset and trains a model to predict the running time of a particular model based on the size and the features of the dataset. Simply, a new dataset is considered as a new row in the error matrix. In order to find the best machine learning algorithm for a new dataset, OBOE runs a particular set of models corresponding to a subset of columns in the error matrix which are predicted to run efficiently on the new dataset. In order to find the rest of the entries in the row, the performance of the models that have not been evaluated are predicted. The good thing about this approach is that it infers the performance of lots of models without the need to run them or even computing meta-features and that is why OBOE can find a well performing model within a reasonable time budget.

The PMF[14] AutoML framework is based on collaborative filtering and Bayesian optimization [66]. More specifically, the problem of selecting the best performing pipeline for a specific task is modeled as a collaborative filtering problem that is solved using probabilistic matrix factorization techniques. PMF considers two datasets to be similar if they have similar evaluations on a few set of pipelines and hence it is more likely that these datasets will have similar evaluations on the rest of the pipelines. This concept is quite related to collaborative filtering for movie recommendation in which users that had the same preference in the past are more likely to have the same preference in the future. In particular, the PMF framework trains each machine learning pipeline on a sample of each dataset and then evaluates such pipeline. This results in a matrix that summarizes the performance (accuracy or balanced accuracy for classification tasks and RMSE for regression tasks) of each machine learning pipeline of each dataset. The problem of predicting the performance of a particular pipeline on a new dataset can be mapped into a matrix factorization problem.

VDS [67] has been recently introduced as an *interactive* automated machine learning tool, that followed the ideas of a previous work on the MLBase

---

[13] https://github.com/udellgroup/oboe/tree/master/automl.
[14] https://github.com/rsheth80/pmf-automl.

framework [68]. In particular, it uses a meta learning mechanism (knowledge from the previous runs) to provide the user with a quick feedback, in few seconds, with an initial model recommendation that can achieve a reasonable accuracy while, on the back-end, conducting an optimization process so that it can recommend to the user more models with better accuracies, as it progresses with the search process over the search space. The VDS framework combines cost-based Multi-Armed Bandits and Bayesian optimizations for exploring the search space while using a rule-based search-space as query optimization technique. VDS prunes unpromising pipelines in early stages using an adaptive pipeline selection algorithm. In addition, it supports a wide range of machine learning tasks including classification, regression, community detection, graph matching, image classification, and collaborative filtering. *ATMSeer*[15] is an interactive visualization tool that has been introduced to support users for refining the search space of AutoML and analyzing the results [69]. Table 1 shows a summary of the main features of the centralized state-of-the-art AutoML frameworks.

Several cloud-based solutions have been introduced to tackle the automated machine learning problem using the availability of high computational power on cloud environments to try a wide range of models and configurations. For example, *Google AutoML*[16] supports training a wide range of machine learning models in different domains with minimal user experience. *Azure AutoML*[17] is a cloud-based service that can be used to automate building machine learning pipeline for both classification and regression tasks. AutoML Azure uses collaborative filtering and Bayesian optimization to search for the most promising pipelines efficiently [66] based on a database that is constructed by running millions of experiments of evaluation of different pipelines on many datasets. *Amazon Sage Maker*[18] provides its users with a wide set of most popular machine learning, and deep learning frameworks to build their models in addition to automatic tuning for the model parameters.

Distributed Frameworks. As the size of the dataset increases, solving the *CASH* problem in a centralized manner turns out to be infeasible due to the limited computing resources (e.g., Memory, CPU) of a single machine. Thus, there is a clear need for distributed solutions that can harness the power of computing clusters that have multiple nodes to tackle the computational complexity of the problem. *MLbase*[19] has been the first work to introduce the idea of developing a distributed framework of machine learning algorithm selection and hyperparameter optimization [68]. MLbase has been based on `MLlib` [70], a

---

[15] https://github.com/HDI-Project/ATMSeer.
[16] https://cloud.google.com/automl/.
[17] https://docs.microsoft.com/en-us/azure/machine-learning/service/.
[18] https://aws.amazon.com/machine-learning/.
[19] http://www.mlbase.org/.

Table 1. Summary of the main features of centralized AutoML frameworks

	Release date	Core language	Training framework	Optimization technique	ML task	Meta learning	User interface	Automatic feature extraction	Open source
AutoWeka	2013	Java	Weka	Bayesian optimization	Single-label classification regression	✗	✓	✓	✓
AutoSklearn	2015	Python	Scikit-learn,	Bayesian optimization	Single-label classification regression	✓	✗	✓	✓
TPOT	2016	Python	Scikit-learn	Genetic algorithm	Single-label classification regression	✗	✗		✓
SmartML	2019	R	mlr, RWeka & Other R packages	Bayesian optimization	Single-label classification	✓	✓	✗	✓
Auto-MEKA$_{GGP}$	2018	Java	Meka	Grammar-based genetic algorithm	Multi-label classification	✓	✗	✗	✓
Recipe	2017	Python	Scikit-learn	Grammar-based genetic algorithm	Single-label classification	✓	✗	✓	✓
MLPlan	2018	Java	Weka and Scikit-learn	Hierachical task planning	Single-label classification	✗	✗	✓	✓
Hyperopt-sklearn	2014	Python	Scikit-learn	Bayesian optimization& Random search	Single-label classification regression	✗	✗	✓	✓
Autostacker	2018	–	–	Genetic algorithm	Single-label classification	✗	✗	✓	✗
VDS	2019	–	–	Cost-based multi-armed bandits and Bayesian optimization	Single-label classification regression image classification audio classification graph matching	✓	✓	✓	✗
AlphaD3M	2018	–	–	Reinforcement learning	Single-label classification regression	✓	✗	✓	✗
OBOE	2019	Python	Scikit-learn	Collaborative filtering	Single-label classification	✓	✗	✗	✓
PMF	2018	Python	Scikit-learn	Collaborative filtering & Bayesian optimization	Single-label classification	✓	✗	✓	✓

Spark-based ML library. It attempted to reused cost-based query optimization techniques to prune the search space at the level of *logical learning plan* before transforming it into a *physical learning plan* to be executed.

Auto-Tuned Models (ATM) framework[20] has been introduced as a parallel framework for fast optimization of machine learning modeling pipelines [71]. In particular, this framework depends on parallel execution along multiple nodes with a shared model hub that stores the results out of these executions and tries to enhance the selection of other pipelines that can outperform the current chosen ones. The user can decide to use either of ATM's two searching methods, a hybrid Bayesian and multi-armed bandit optimization system, or a model recommendation system that works by exploiting the previous performance of modeling techniques on a variety of datasets.

TransmogrifAI[21] is one of the most recent modular tools written in Scala. It is built using workflows of feature preprocessors, and model selectors on top of Spark with minimal human involvement. It has the ability to reuse the selected work-flows. Currently, `TransmogrifAI` supports eight different binary classifiers and five regression algorithms. *MLBox*[22] is a Python-based AutoML framework for distributed preprocessing, optimization and prediction. MLBox supports model stacking where a new model is trained from the combined predictors of multiple previously trained models. It uses `hyperopt`[23], a distributed asynchronous hyper-parameter optimization library, in Python, to perform the hyper-parameter optimisation process.

Rafiki[24] has been introduced as a distributed framework which is based on the idea of using previous models that achieved high performance on the same tasks [72]. In this framework, regarding the data and parameter storage, the data uploaded by user to be trained is stored in a Hadoop Distributed File System (HDFS). During training, there is a database for each model storing the best version of parameters from hyper-parameter tuning process. This database is kept in memory as it is accessed and updated frequently. Once the hyper-parameter tuning process is finished, the database is dumped to the disk. The types of parameters to be tuned are either related to model architecture like number of Layers, and Kernel or related to the training algorithm itself like weight decay, and learning rate. All these parameters can be tuned using a random search or Bayesian optimization. Table 2 shows a summary of the main features of the distributed AutoML frameworks.

---

[20] https://github.com/HDI-Project/ATM.
[21] https://transmogrif.ai/.
[22] https://github.com/AxeldeRomblay/MLBox.
[23] https://github.com/hyperopt/hyperopt.
[24] https://github.com/nginyc/rafiki.

Table 2. Summary of the main features of distributed AutoML frameworks

	Release date	Core language	Optimization technique	Training framework	Meta-learning	User interface	Open source
MLBase	2013	Scala	Cost-based multi-armed bandits	Spark MLlib	×	×	×
ATM	2017	Python	Hybrid Bayesian, and multi-armed bandits optimization	Scikit-learn	✓	×	✓
MLBox	2017	Python	Distributed random search, Tree-Parzen estimators	Scikit-learn Keras	×	×	✓
Rafiki	2018	Python	Distributed random search, Bayesian optimization	TensorFlow Scikit-learn	×	✓	✓
TransmogrifAI	2018	Scala	Bayesian optimization, and random search	SparkML	×	×	✓

Table 3. Summary of the main features of the neural architecture search frameworks

	Release date	Open source	Optimization technique	Supported frameworks	Interface
Auto Keras	2018	✓	Network morphism	Keras	✓
Auto Net	2016	✓	SMAC	PyTorch	✗
NNI	2019	✓	Random search different Bayesian optimizations annealing network morphism hyper-band naive evolution grid search	PyTorch, TensorFlow, Keras, Caffe2, CNTK, Chainer Theano	✓
enas	2018	✓	Reinforcement learning	Tensorflow	✗
NAO	2018	✓	Gradient based optimization	Tensorflow, PyTorch	✗
DARTS	2019	✓	Gradient based optimization	PyTorch	✗
LEAF	2019	✗	Evolutionary algorithms	–	✗

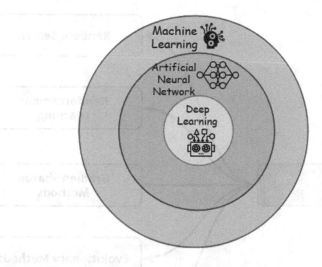

Fig. 5. The relationship between machine learning and deep learning.

3   Automated Deep Learning

3.1   Neural Architecture Search for Deep Learning

In general, deep learning techniques [73] represent a subset of machine learning methodologies that are based on artificial neural networks (ANN) which
are mainly inspired by the neuron structure of the human brain (Fig. 5). It is
described as *deep* because it has more than one layer of nonlinear feature transformation. Neural Architecture Search (NAS) is a fundamental step in automating the machine learning process and has been successfully used to design the
model architecture for image and language tasks [74–78]. Broadly, NAS techniques falls into five main categories including *random search, reinforcement
learning, gradient-based methods, evolutionary methods*, and *Bayesian optimization* (Fig. 6).

Random search is one of the most naive and simplest approaches for network
architecture search. For example, Hoffer et al. [79] have presented an approach
to find good network architecture using a random search combined with well-
trained set of shared weights. Li and Talwalkar [80] proposed new network architecture search baselines that are based on a random search with early-stopping
for hyper-parameter optimization. Results show that random search along with
early-stopping achieves the state-of-the-art network architecture search results
on two standard NAS bookmarkers which are PTB and CIFAR-10 datasets.

Reinforcement learning [81] is another approach that has been used to find
the best network architecture. Zoph and Le [74] used a recurrent neural network (LSTM) with reinforcement to compose neural network architecture. More
specifically, recurrent neural network is trained through a gradient based search
algorithm called REINFORCE [82] to maximize the expected accuracy of the generated neural network architecture. Baker et al. [83] introduced a meta-modeling

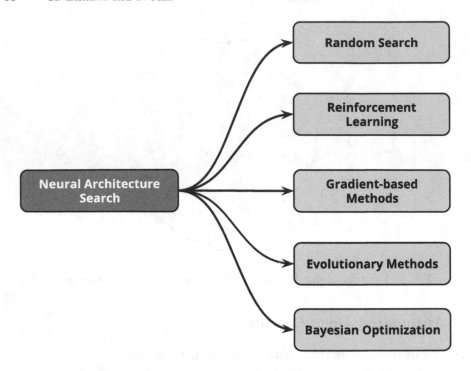

Fig. 6. A taxonomy for the Neural Network Architecture Search (NAS) techniques

algorithm called `MetaQNN` based on reinforcement learning to automatically generate the architecture of a convolutional neural network for a new task. The convolutional neural network layers are chosen sequentially by a learning agent that is trained using $Q-$learning with $\epsilon-$greedy exploration technique. Simply, the agent explores a finite search space of a set of architectures and iteratively figures out architecture designs with improved performance on the new task to be learned.

Gradient-based optimization is another common way for neural network architecture search. Liu et al. [84] proposed an approach based on continuous relaxation of the neural architecture allowing using a gradient descent for architecture search. Experiments showed that this approach excels in finding high-performance convolutional architectures for image classification tasks on CIFAR-10, and ImageNet datasets. Shin et al. [85] proposed a gradient-based optimization approach for learning the network architecture and parameters simultaneously. Ahmed and Torresani [86] used gradient based approach to learn network architecture. Experimental results on two different networks architecture ResNet and ResNeXt show that this approach yields to better accuracy and a significant reduction in the number of parameters.

Another direction for architecture search is *evolutionary algorithms* which are well suited for optimizing arbitrary structure. Miller et al. [87] considered an evolutionary algorithm to propose the architecture of the neural network and

network weights as well. Many evolutionary approaches based on genetic algorithms are used to optimize the neural networks architecture and weights [88–90] while others rely on hierarchical evolution [78]. Some recent approaches consider using the multi-objective evolutionary architecture search to optimize training time, complexity and performance [91,92] of the network. LEAF [93] is an evolutionary AutoML framework that optimizes hyper-parameters, network architecture and the size of the network. LEAF uses CoDeepNEAT [94] which is a powerful evolutionary algorithm based on NEAT [95]. LEAF achieved the state-of-the-art performance results on medical image classification and natural language analysis. For supervised learning tasks, evolutionary based approaches tend to outperform reinforcement learning approaches especially when the neural network architecture is very complex due to having millions of parameters to be tuned. For example, the best performance achieved on ImageNet and CIFAR-10 has been obtained using evolutionary techniques [96].

Bayesian optimization based on Gaussian processes has been used by Kandasamy et al. [97] and Swersky et al. [98] for tackling the neural architecture search problem. In addition, lots of work focused on using tree based models such as random forests and tree Parzen estimators [44] to effectively optimize the network architecture as well as its hyper-parameters [45,52,99]. Bayesian optimization may outperform evolutionary algorithms in some problems as well [100].

3.2   AutoDL Frameworks

Recently, some frameworks (e.g., Auto-Keras [101], and Auto-Net [99]) have been proposed with the aim of automatically finding neural network architectures that are competitive with architectures designed by human experts. However, the results so far are not significant. For example, *Auto-Keras* [101] is an open source efficient neural architecture search framework based on Bayesian optimization to guide the network morphism. In order to explore the search space efficiently, Auto-Keras uses a neural network kernel and tree structured acquisition function with iterative Bayesian optimization. First, a Gaussian process model is trained on the currently existing network architectures and their performance is recorded. Then, the next neural network architecture obtained by the acquisition function is generated and evaluated. Moreover, Auto-Keras runs in a parallel mode on both CPU and GPU.

Auto-Net [99] is an efficient neural architecture search framework based on SMAC optimization and built on top of PyTorch. The first version of Auto-Net is implemented within the Auto-sklearn in order to leverage some of the existing components of the machine learning pipeline in Auto-sklearn such as preprocessing. The first version of Auto Net only considers fully-connected feed-forward neural networks as they are applied on a large number of different datasets. Auto-net accesses deep learning techniques from Lasagne Python deep learning library [102]. Auto Net includes a number of algorithms for tuning the neural network weights including vanilla stochastic gradient descent, stochastic gradient descent with momentum, Adadelta [103], Adam [104], Nesterov momentum [105] and Adagrad [106].

Neural Network Intelligence (NNI)[25] is an open source toolkit by Microsoft that is used for tuning neural networks architecture and hyper-parameters in different environments including local machine, cloud and remote servers. NNI accelerates and simplifies the huge search space using built-in super-parameter selection algorithms including random search, naive evolutionary algorithms, simulated annealing, network morphism, grid search, hyper-band, and a bunch of Bayesian optimizations like SMAC [43], and BOHB [47]. NNI supports a large number of deep leaning frameworks including PyTorch, TensorFlow, Keras, Caffe2, CNTK, Chainer and Theano.

DEvol [26] is an open source framework for neural network architecture search that is based on genetic programming to evolve the number of layers, kernels, and filters, the activation function and dropout rate. DEvol uses parallel training in which multiple members of the population are evaluated across multiple GPU machines in order to accelerate the process of finding the most promising network.

enas [107] has been introduced as an open source framework for neural architecture search in Tensorflow based on reinforcement learning [74] where a controller of a recurrent neural network architecture is trained to search for optimal subgraphs from large computational graphs using policy gradient. Moreover, *enas* showed a large speed up in terms of GPU hours thanks to the sharing of parameters across child subgraphs during the search process.

NAO [108], and *Darts* [84] are open source frameworks for neural architecture search which propose a new continuous optimization algorithm that deals with the network architecture as a continuous space instead of the discretization followed by other approaches. In *NAO*, the search process starts by encoding an initial architecture to a continuous space. Then, a performance predictor based on gradient based optimization searches for a better architecture that is decoded at the end by a complementary algorithm to the encoder in order to map the continuous space found back into its architecture. On the other hand, *DARTS* learns new architectures with complex graph topologies from the rich continuous search space using a novel bilevel optimization algorithm. In addition, it can be applied to any specific architecture family without restrictions to any of the convolutional and recurrent networks only. Both frameworks showed a competitive performance using limited computational resources compared with other neural architecture search frameworks.

Evolutionary Neural AutoML for Deep Learning (LEAF) [93] is an AutoML framework that optimizes neural network architecture and hyper-parameters using the state-of-the-art evolutionary algorithm and distributed computing framework. LEAF uses CoDeepNEAT [94] for optimizing deep neural network architecture and hyper-parameters. LEAF consists of three main layers which are algorithm layers, system layer and problem-domain layer. LEAF evolves deep neural networks architecture and hyper-parameters in the algorithm layer. The system layer is responsible for training the deep neural networks in a parallel

---

[25] https://github.com/Microsoft/nni.
[26] https://github.com/joeddav/devol.

mode on a cloud environment such as Microsoft Azure[27], Google Cloud[28] and Amazon AWS[29], which is essential in the evaluation of the fitness of the neural networks evolved in the algorithm layer. More specifically, the algorithm layer sends the neural network architecture to the system layer. Then, the system layer sends the evaluation of the fineness of this network back to the algorithm layer. Both the algorithm layer and the system layer work together to support the problem-domain layers where the problems of hyper-parameter tuning of network architecture search are solved. Table 3 shows a summary of the main features of the state-of-the-art neural architecture search frameworks.

4 Open Challenges and Future Directions

Although in the last years, there has been increasing research efforts to tackle the challenges of the automated machine learning domain, however, there are still several open challenges and research directions that needs to be tackled to achieve the ultimate goals and vision of the AutoML domain. In this section, we highlight some of these challenges that need to be tackled to improve the state-of-the-art.

Scalability: In practice, a main limitation of the centralized frameworks for automating the solutions for the CASH problem (e.g., `Auto-Weka`, `Auto-Sklearn`) is that they are tightly coupled with a machine learning library (e.g., `Weka`, `scikit-learn`, R) that can only work on a *single* node which makes them not applicable in the case of large data volumes. In practice, as the scale of data produced daily is increasing continuously at an exponential scale, several distributed machine learning platforms have been recently introduced. Examples include `Spark MLib` [70], `Mahout`[30] and `SystemML` [109]. Although there have been some initial efforts for distributed automated framework for the CASH problem. However, the proposed distributed solutions are still simple and limited in their capabilities. More research efforts and novel solutions are required to tackle the challenge of automatically building and tuning machine learning models over massive datasets.

Optimization Techniques: In practice, different AutoML frameworks use different techniques for hyper-parameter optimization of the machine learning algorithms. For instance, `Auto-Weka` and `Auto-Sklearn` use the SMAC technique with cross-fold validation during the hyper-parameter configuration optimization and evaluation. On the other hand, `ML-Plan` uses the hierarchical task network with Monte Carlo Cross-Validation. Other tools, including `Recipe` [61] and `TPOT`, use genetic programming, and pareto optimization for generating candidate pipelines. In practice, it is difficult to find a clear winner or one-size-fits-all

---

[27] https://azure.microsoft.com/en-us/.
[28] https://cloud.google.com/.
[29] https://aws.amazon.com/.
[30] https://mahout.apache.org/.

technique. In other words, there is no single method that will be able to outperform all other techniques on the different datasets with their various characteristics, types of search spaces and metrics (e.g., time and accuracy). Thus, there is a crucial need to understand the Pros and Cons of these optimization techniques so that AutoML systems can automatically tune their hyper-parameter optimization techniques or their strategy for exploring and traversing the search space. Such decision automation should provide improved performance over picking and relying on a fixed strategy. Similarly, for the various introduced meta-learning techniques, there is no clear systematic process or evaluation metrics to quantitatively assess and compare the impact of these techniques on reducing the search space. Recently, some competitions and challenges[31,32] have been introduced and organized to address this issue such as the DARPA D3M Automatic Machine Learning competition [67].

Time Budget: A common important parameter for AutoML systems is the user *time budget* to wait before getting the recommended pipeline. Clearly, the bigger the time budget, the more the chance for the AutoML system to explore various options in the search space and the higher probability to get a better recommendation. However, the bigger time budget used, the longer waiting time and the higher computing resource consumption, which could be translated into a higher monetary bill in the case of using cloud-based resources. On the other hand, a small-time budget means a shorter waiting time but a lower chance to get the best recommendation. However, it should be noted that increasing the time budget from X to $2X$ does not necessarily lead to a big increase on the quality of the results of the recommended pipeline, if any at all. In many scenarios, this extra time budget can be used for exploring more of the unpromising branches in the search space or exploring branches that have very little gain, if any. For example, the accuracy of the returned models from running the `AutoSklearn` framework over the `Abalone` dataset[33] with time budgets of 4 h and 8 h are almost the same (25%). Thus, accurately estimating or determining the adequate time budget to optimize this trade-off is another challenging decision that can not be done by non-expert end users. Therefore, it is crucial to tackle such challenge by automatically predicting/recommending the adequate time budget for the modeling process. The `VDS` [67] framework provided a first attempt to tackle this challenge by proposing an interactive approach that relies on meta learning to provide a quick first model recommendation that can achieve a reasonable quality while conducting an offline optimization process and providing the user with a *stream* of models with better accuracy. However, more research efforts to tackle this challenge are still required.

Composability. Nowadays, several machine learning solutions (e.g., `Weka`, `Scikit-Learn`, `R`, `MLib`, `Mahout`) have become popular. However, these ML solutions significantly vary in their available techniques (e.g., learning algorithms,

---

[31] https://www.4paradigm.com/competition/nips2018.
[32] http://automl.chalearn.org/.
[33] https://www.openml.org/d/183.

preprocessors, and feature selectors) to support each phase of the machine learning pipeline. Clearly, the quality of the machine learning pipelines that can be produced by any of these platforms depends on the availability of several techniques/algorithms that can be utilized in each step of the pipeline. In particular, the more available techniques/algorithms in a machine learning platform, the higher the ability and probability of producing a well-performing machine learning pipeline. In practice, it is very challenging to have optimized implementations for all of the algorithms/techniques of the different steps of the machine learning pipeline available in a single package, or library. The ML-Plan framework [62] has been attempting to tackle the composability challenge on building machine learning pipelines. In particular, it integrates a superset of both Weka and Scikit-Learn algorithms to construct a full pipeline. The initial results of this approach have shown that the composable pipelines over Weka and Scikit-Learn do not significantly outperform the outcomes from Auto-Weka and Auto-Sklearn frameworks especially with big datasets and small time budgets. However, we believe that there are several reasons behind these results. First, combining the algorithms/techniques of more than one machine learning platform causes a dramatic increase in the search space. Thus, to tackle this challenge, there is a crucial need for a smart and efficient search algorithm that can effectively reduce the search space and focus on the promising branches. Using meta-learning approaches can be an effective solution to tackle this challenge. Second, combining services from more than one framework can involve a significant overhead for the data and message communications between the different frameworks. Therefore, there is a crucial need for a smart *cost-based* optimizer that can accurately estimate the gain and cost of each recommended composed pipeline and be able to choose the composable recommendations when they are able to achieve a clear performance gain. Third, the ML-Plan has been combining the services of two single node machine learning services (Weka and Scikit-Learn). We believe that the best gain of the composability mechanism will be achieved by combining the performance power of distributed systems (e.g., MLib) with the rich functionality of many centralized systems.

User Friendliness: In general, most of the current tools and framework can not be considered to be user friendly. They still need sophisticated technical skills to be deployed and used. Such challenge limits its usability and wide acceptance among layman users and domain experts (e.g., physicians, accountants) who commonly have limited technical skills. Providing an interactive and light-weight web interfaces for such framework can be one of the approaches to tackle these challenges.

Continuous Delivery Pipeline: Continuous delivery is defined as creating a repeatable, reliable and incrementally improving process for taking software from concept to customer. Integrating machine learning models into continuous delivery pipelines for productive use has not recently drawn much attention, because usually the data scientists push them directly into the production environment with all the drawbacks this approach may have, such as no proper unit and integration testing.

5   Conclusion

Machine learning has become one of the main engines of the current era. The production pipeline of a machine learning models passe through different phases and stages that require a wide knowledge of several available tools, and algorithms. However, as the scale of data produced daily is increasing continuously at an exponential scale, it has become essential to automate this process. In this chapter, we provided an overview of the state-of-the-art research effort in the domain of AutoML frameworks. We have also highlighted research directions and open challenges that need to be addressed in order to achieve the vision and goals of the AutoML process. We hope that our overview serves as a useful resource for the community, for both researchers and practitioners, to understand the challenges of the domain and provide useful insight for further advancing the state-of-the-art in several directions.

Acknowledgment. This work of Sherif Sakr is funded by the European Regional Development Funds via the Mobilitas Plus programme (grant MOBTT75). The work of Radwa Elshawi is funded by the European Regional Development Funds via the Mobilitas Plus programme (MOBJD341). The authors would like to thank Mohamed Maher for his comments.

References

1. Zomaya, A.Y., Sakr, S. (eds.): Handbook of Big Data Technologies. Springer, Cham (2017). https://doi.org/10.1007/978-3-319-49340-4
2. Sakr, S., Zomaya, A.Y. (eds.): Encyclopedia of Big Data Technologies. Springer, Cham (2019). https://doi.org/10.1007/978-3-319-77525-8
3. Fitzgerald, B.: Software crisis 2.0. Computer 45(4), 89–91 (2012)
4. Vafeiadis, T., Diamantaras, K.I., Sarigiannidis, G., Chatzisavvas, K.C.: A comparison of machine learning techniques for customer churn prediction. Simul. Modell. Pract. Theory 55, 1–9 (2015)
5. Probst, P., Boulesteix, A.-L.: To tune or not to tune the number of trees in random forest. J. Mach. Learn. Res. 18, 181–1 (2017)
6. Pedregosa, F., et al.: Scikit-learn: machine learning in python. J. Mach. Learn. Res. 12, 2825–2830 (2011)
7. Kotthoff, L., Thornton, C., Hoos, H.H., Hutter, F., Leyton-Brown, K.: Auto-WEKA 2.0: automatic model selection and hyperparameter optimization in WEKA. J. Mach. Learn. Res. 18(1), 826–830 (2017)
8. Feurer, M., Klein, A., Eggensperger, K., Springenberg, J.T., Blum, M., Hutter, F.: Efficient and robust automated machine learning. In Proceedings of the 28th International Conference on Neural Information Processing Systems, NIPS 2015, vol. 2, pp. 2755–2763 (2015). MIT Press, Cambridge
9. Maher, M., Sakr, S.: SmartML: a meta learning-based framework for automated selection and hyperparameter tuning for machine learning algorithms. In EDBT: 22nd International Conference on Extending Database Technology (2019)
10. Brazdil, P., Carrier, C.G., Soares, C., Vilalta, R.: Metalearning: Applications to Data Mining. Springer, Heidelberg (2008). https://doi.org/10.1007/978-3-540-73263-1

11. Vanschoren, J.: Meta-learning: a survey. CoRR, abs/1810.03548 (2018)
12. Bilalli, B., Abelló, A., Aluja-Banet, T.: On the predictive power of meta-features in OpenML. Int. J. Appl. Math. Comput. Sci. **27**(4), 697–712 (2017)
13. Bardenet, R., Brendel, M., Kégl, B., Sebag, M.: Collaborative hyperparameter tuning. In: International Conference on Machine Learning, pp. 199–207 (2013)
14. Soares, C., Brazdil, P.B., Kuba, P.: A meta-learning method to select the kernel width in support vector regression. Mach. Learn. **54**(3), 195–209 (2004)
15. Nisioti, E., Chatzidimitriou, K., Symeonidis, A.: Predicting hyperparameters from meta-features in binary classification problems. In: AutoML Workshop at ICML (2018)
16. Köpf, C., Iglezakis, I.: Combination of task description strategies and case base properties for meta-learning. In: Proceedings of the 2nd International Workshop on Integration and Collaboration Aspects of Data Mining, Decision Support and Meta-learning, pp. 65–76 (2002)
17. Krizhevsky, A., Sutskever, I., Hinton, G.E.: ImageNet classification with deep convolutional neural networks. In: Advances in Neural Information Processing Systems, pp. 1097–1105 (2012)
18. Giraud-Carrier, C.: Metalearning-a tutorial. In: Tutorial at the 7th International Conference on Machine Learning and Applications (ICMLA), San Diego, California, USA (2008)
19. Brazdil, P.B., Soares, C., Da Costa, J.P.: Ranking learning algorithms: using IBL and meta-learning on accuracy and time results. Mach. Learn. **50**(3), 251–277 (2003)
20. dos Santos, P.M., Ludermir, T.B., Prudencio, R.B.C.: Selection of time series forecasting models based on performance information. In: Fourth International Conference on Hybrid Intelligent Systems (HIS 2004), pp. 366–371. IEEE (2004)
21. Reif, M., Shafait, F., Goldstein, M., Breuel, T., Dengel, A.: Automatic classifier selection for non-experts. Pattern Anal. Appl. **17**(1), 83–96 (2014)
22. Guerra, S.B., Prudêncio, R.B.C., Ludermir, T.B.: Predicting the performance of learning algorithms using support vector machines as meta-regressors. In: Kůrková, V., Neruda, R., Koutník, J. (eds.) ICANN 2008. LNCS, vol. 5163, pp. 523–532. Springer, Heidelberg (2008). https://doi.org/10.1007/978-3-540-87536-9_54
23. Pan, S.J., Yang, Q.: A survey on transfer learning. IEEE Trans. Knowl. Data Eng. **22**(10), 1345–1359 (2010)
24. Bengio, Y.: Deep learning of representations for unsupervised and transfer learning. In: Proceedings of ICML Workshop on Unsupervised and Transfer Learning, pp. 17–36 (2012)
25. Baxter, J.: Learning internal representations. Flinders University of South Australia (1995)
26. Caruana, R.: Learning many related tasks at the same time with backpropagation. In: Advances in Neural Information Processing Systems, pp. 657–664 (1995)
27. Razavian, A.S., Azizpour, H., Sullivan, J., Carlsson, S.: CNN features off-the-shelf: an astounding baseline for recognition. In: Proceedings of the IEEE Conference on Computer Vision and Pattern Recognition Workshops, pp. 806–813 (2014)
28. Donahue, J., et al.: Decaf: a deep convolutional activation feature for generic visual recognition. In: International Conference on Machine Learning, pp. 647–655 (2014)
29. Yosinski, J., Clune, J., Bengio, Y., Lipson, H.: How transferable are features in deep neural networks? In: Advances in Neural Information Processing Systems, pp. 3320–3328 (2014)

30. Davis, L.: Handbook of genetic algorithms. In: Glover, F., Kochenberger, G.A. (eds.) Handbook of Metaheuristics. International Series in Operations Research & Management Science. Springer, Boston (1991)
31. Pelikan, M., Goldberg, D.E., Cantú-Paz, E.: Boa: the Bayesian optimization algorithm. In: Proceedings of the 1st Annual Conference on Genetic and Evolutionary Computation, vol. 1, pp. 525–532. Morgan Kaufmann Publishers Inc. (1999)
32. Polak, E.: Optimization: Algorithms and Consistent Approximations, vol. 124. Springer, New York (2012). https://doi.org/10.1007/978-1-4612-0663-7
33. Montgomery, D.C.: Design and Analysis of Experiments. Wiley, New York (2017)
34. Bergstra, J., Bengio, Y.: Random search for hyper-parameter optimization. J. Mach. Learn. Res. **13**, 281–305 (2012)
35. Zhilinskas, A.G.: Single-step Bayesian search method for an extremum of functions of a single variable. Cybern. Syst. Anal. **11**(1), 160–166 (1975)
36. Jones, D.R., Schonlau, M., Welch, W.J.: Efficient global optimization of expensive black-box functions. J. Global Optim. **13**(4), 455–492 (1998)
37. Snoek, J., et al.: Scalable Bayesian optimization using deep neural networks. In: International Conference on Machine Learning, pp. 2171–2180 (2015)
38. Snoek, J., Larochelle, H., Adams, R.P.: Practical Bayesian optimization of machine learning algorithms. In: Advances in Neural Information Processing Systems, pp. 2951–2959 (2012)
39. Dahl, G.E., Sainath, T.N., Hinton, G.E.: Improving deep neural networks for LVCSR using rectified linear units and dropout. In: IEEE International Conference on Acoustics, Speech and Signal Processing, pp. 8609–8613. IEEE (2013)
40. Melis, G., Dyer, C., Blunsom, P.: On the state of the art of evaluation in neural language models. arXiv preprint arXiv:1707.05589 (2017)
41. Martinez-Cantin, R.: BayesOpt: a Bayesian optimization library for nonlinear optimization, experimental design and bandits. J. Mach. Learn. Res. **15**(1), 3735–3739 (2014)
42. Breiman, L.: Random forests. Mach. Learn. **45**(1), 5–32 (2001)
43. Hutter, F., Hoos, H.H., Leyton-Brown, K.: Sequential model-based optimization for general algorithm configuration. In: Coello, C.A.C. (ed.) LION 2011. LNCS, vol. 6683, pp. 507–523. Springer, Heidelberg (2011). https://doi.org/10.1007/978-3-642-25566-3_40
44. Bergstra, J.S., Bardenet, R., Bengio, Y., Kégl, B.: Algorithms for hyper-parameter optimization. In: Advances in Neural Information Processing Systems, pp. 2546–2554 (2011)
45. Bergstra, J., Yamins, D., Cox, D.D.: Making a science of model search: hyperparameter optimization in hundreds of dimensions for vision architectures (2013)
46. Eggensperger, K., et al.: Towards an empirical foundation for assessing Bayesian optimization of hyperparameters. In: NIPS Workshop on Bayesian Optimization in Theory and Practice, vol. 10, p. 3 (2013)
47. Falkner, S., Klein, A., Hutter, F.: BOHB: robust and efficient hyperparameter optimization at scale. arXiv preprint arXiv:1807.01774 (2018)
48. Sparks, E.R., Talwalkar, A., Haas, D., Franklin, M.J., Jordan, M.I., Kraska, T.: Automating model search for large scale machine learning. In: Proceedings of the Sixth ACM Symposium on Cloud Computing, pp. 368–380. ACM (2015)
49. Kirkpatrick, S., Gelatt, C.D., Vecchi, M.P.: Optimization by simulated annealing. Science **220**(4598), 671–680 (1983)
50. Holland, J.H., et al.: Adaptation in Natural and Artificial Systems: an Introductory Analysis with Applications to Biology, Control, and Artificial Intelligence. MIT press, Cambridge (1992)

51. Fernández-Godino, M.G., Park, C., Kim, N.-H., Haftka, R.T.: Review of multi-fidelity models. arXiv preprint arXiv:1609.07196 (2016)
52. Domhan, T., Springenberg, J.T., Hutter, F.: Speeding up automatic hyperparameter optimization of deep neural networks by extrapolation of learning curves. In: Twenty-Fourth International Joint Conference on Artificial Intelligence (2015)
53. Jamieson, K.G., Talwalkar, A.: Non-stochastic best arm identification and hyperparameter optimization. In: AISTATS, pp. 240–248 (2016)
54. Li, L., Jamieson, K., DeSalvo, G., Rostamizadeh, A., Talwalkar, A.: Hyperband: a novel bandit-based approach to hyperparameter optimization. arXiv preprint arXiv:1603.06560 (2016)
55. de Sá, A.G.C., Freitas, A.A., Pappa, G.L.: Automated selection and configuration of multi-label classification algorithms with grammar-based genetic programming. In: Auger, A., Fonseca, C.M., Lourenço, N., Machado, P., Paquete, L., Whitley, D. (eds.) PPSN 2018. LNCS, vol. 11102, pp. 308–320. Springer, Cham (2018). https://doi.org/10.1007/978-3-319-99259-4_25
56. Tsoumakas, G., Katakis, I., Vlahavas, I.: Mining multi-label data. In: Maimon, O., Rokach, L. (eds.) Data Mining and Knowledge Discovery Handbook, pp. 667–685. Springer, Boston (2010). https://doi.org/10.1007/978-0-387-09823-4_34
57. Read, J., Reutemann, P., Pfahringer, B., Holmes, G.: MEKA: a multi-label/multi-target extension to WEKA. J. Mach. Learn. Res. **17**(1), 667–671 (2016)
58. Komer, B., Bergstra, J., Eliasmith, C.: Hyperopt-sklearn: automatic hyperparameter configuration for scikit-learn. In: ICML Workshop on AutoML, pp. 2825–2830 (2014)
59. Bergstra, J.., Yamins, D., Cox, D.D.: Hyperopt: a python library for optimizing the hyperparameters of machine learning algorithms. In: Proceedings of the 12th Python in Science Conference, pp. 13–20 (2013)
60. Olson, R.S., Moore, J.H.: TPOT:: a tree-based pipeline optimization tool for automating machine learning. In: Hutter, F., Kotthoff, L., Vanschoren, J. (eds.) Proceedings of the Workshop on Automatic Machine Learning, volume 64 of Proceedings of Machine Learning Research, pp. 66–74, New York, USA, 24 Jun 2016. PMLR
61. de Sá, A.G.C., Pinto, W.J.G.S., Oliveira, L.O.V.B., Pappa, G.L.: RECIPE: a grammar-based framework for automatically evolving classification pipelines. In: McDermott, J., Castelli, M., Sekanina, L., Haasdijk, E., García-Sánchez, P. (eds.) EuroGP 2017. LNCS, vol. 10196, pp. 246–261. Springer, Cham (2017). https://doi.org/10.1007/978-3-319-55696-3_16
62. Mohr, F., Wever, M., Hüllermeier, E.: ML-plan: automated machine learning via hierarchical planning. Mach. Learn. **107**(8–10), 1495–1515 (2018)
63. Chen, B., Wu, H., Mo, W., Chattopadhyay, I., Lipson, H.: Autostacker: a compositional evolutionary learning system. In: Proceedings of the Genetic and Evolutionary Computation Conference, GECCO 2018, pp. 402–409. ACM, New York (2018)
64. Drori, I., et al.: AlphaD3M: machine learning pipeline synthesis. In: AutoML Workshop at ICML (2018)
65. Yang, C., Akimoto, Y., Kim, D.W., Udell, M.: OBOE: collaborative filtering for AutoML initialization. arXiv preprint arXiv:1808.03233 (2019)
66. Fusi, N., Sheth, R., Elibol, H.M.: Probabilistic matrix factorization for automated machine learning. arXiv preprint arXiv:1705.05355 (2017)
67. Shang, Z., et al.: Democratizing data science through interactive curation of ml pipelines. In: Proceedings of the ACM SIGMOD International Conference on Management of Data (SIGMOD) (2019)

68. Kraska, T., Talwalkar, A., Duchi, J.C., Griffith, R., Franklin, M.J., Jordan, M.I.: MLbase: a distributed machine-learning system. In: CIDR, vol. 1, pp. 1–2 (2013)
69. Wang, Q., et al.: ATMseer: increasing transparency and controllability in automated machine learning. In: Proceedings of the 2019 CHI Conference on Human Factors in Computing Systems, p. 681. ACM (2019)
70. Meng, X., et al.: MLlib: machine learning in apache spark. J. Mach. Learn. Res. **17**(1), 1235–1241 (2016)
71. Swearingen, T., Drevo, W., Cyphers, B., Cuesta-Infante, A., Ross, A., Veeramachaneni, K.: ATM: a distributed, collaborative, scalable system for automated machine learning, pp. 151–162, December 2017
72. Wei Wang, et al.: Rafiki: machine learning as an analytics service system. CoRR, abs/1804.06087 (2018)
73. Bengio, Y., et al.: Learning deep architectures for AI. Foundations Trends® Mach. Learn. **2**(1), 1–127 (2009)
74. Zoph, B., Le, Q.V.: Neural architecture search with reinforcement learning. arXiv preprint arXiv:1611.01578 (2016)
75. Zoph, B., Vasudevan, V., Shlens, J., Le, Q.V.: Learning transferable architectures for scalable image recognition. In: Proceedings of the IEEE Conference on Computer Vision and Pattern Recognition, pp. 8697–8710 (2018)
76. Cai, H., Chen, T., Zhang, W., Yu, Y., Wang, J.: Efficient architecture search by network transformation. In: Thirty-Second AAAI Conference on Artificial Intelligence (2018)
77. Liu, C., et al.: Progressive neural architecture search. In: Proceedings of the European Conference on Computer Vision (ECCV), pp. 19–34 (2018)
78. Liu, H., Simonyan, K., Vinyals, O., Fernando, C., Kavukcuoglu, K.: Hierarchical representations for efficient architecture search. arXiv preprint arXiv:1711.00436 (2017)
79. Hoffer, E., Hubara, I., Soudry, D.: Train longer, generalize better: Closing the generalization gap in large batch training of neural networks. In: Proceedings of the 31st International Conference on Neural Information Processing Systems, NIPS 2017, USA, pp. 1729–1739. Curran Associates Inc. (2017)
80. Li, L., Talwalkar, A.: Random search and reproducibility for neural architecture search (2019)
81. Sutton, R.S., Barto, A.G., et al.: Introduction to Reinforcement Learning, vol. 135. MIT Press, Cambridge (1998)
82. Williams, R.J.: Simple statistical gradient-following algorithms for connectionist reinforcement learning. Mach. Learn. **8**(3–4), 229–256 (1992)
83. Baker, B., Gupta, O., Naik, N., Raskar, R.: Designing neural network architectures using reinforcement learning. arXiv preprint arXiv:1611.02167 (2016)
84. Liu, H., Simonyan, K., Yang, Y.: Darts: differentiable architecture search. arXiv preprint arXiv:1806.09055 (2018)
85. Shin, R., Packer, C., Song, D.: Differentiable neural network architecture search (2018)
86. Ahmed, K., Torresani, L.: MaskConnect: connectivity learning by gradient descent. In: Ferrari, V., Hebert, M., Sminchisescu, C., Weiss, Y. (eds.) ECCV 2018. LNCS, vol. 11209, pp. 362–378. Springer, Cham (2018). https://doi.org/10.1007/978-3-030-01228-1_22
87. Miller, G.F., Todd, P.M., Hegde, S.U.: Designing neural networks using genetic algorithms. In: ICGA, vol. 89, pages 379–384 (1989)
88. Stanley, K.O., Miikkulainen, R.: Evolving neural networks through augmenting topologies. Evol. Comput. **10**(2), 99–127 (2002)

89. Stanley, K.O., D'Ambrosio, D.B., Gauci, J.: A hypercube-based encoding for evolving large-scale neural networks. Artif. Life **15**(2), 185–212 (2009)
90. Angeline, P.J., Saunders, G.M., Pollack, J.B.: An evolutionary algorithm that constructs recurrent neural networks. IEEE Trans. Neural Netw. **5**(1), 54–65 (1994)
91. Lu, Z., et al.: NSGA-Net: a multi-objective genetic algorithm for neural architecture search. arXiv preprint arXiv:1810.03522 (2018)
92. Elsken, T., Metzen, J.H., Hutter, F.: Efficient multi-objective neural architecture search via Lamarckian evolution (2018)
93. Liang, J., Meyerson, E., Hodjat, B., Fink, D., Mutch, K., Miikkulainen, R.: Evolutionary neural AutoML for deep learning (2019)
94. Miikkulainen, R., et al.: Evolving deep neural networks. In: Artificial Intelligence in the Age of Neural Networks and Brain Computing, pp. 293–312. Elsevier (2019)
95. Real, E., et al.: Large-scale evolution of image classifiers. In: Proceedings of the 34th International Conference on Machine Learning, vol. 70, pp. 2902–2911. JMLR. org (2017)
96. Real, E., Aggarwal, A., Huang, Y., Le, Q.V.: Regularized evolution for image classifier architecture search. arXiv preprint arXiv:1802.01548 (2018)
97. Kandasamy, K., Neiswanger, W., Schneider, J., Poczos, B., Xing, E.: Neural architecture search with Bayesian optimisation and optimal transport (2018)
98. Swersky, K., Duvenaud, D., Snoek, J., Hutter, F., Osborne, M.A.: Raiders of the lost architecture: kernels for Bayesian optimization in conditional parameter spaces. arXiv preprint arXiv:1409.4011 (2014)
99. Mendoza, H., Klein, A., Feurer, M., Springenberg, J.T., Hutter, F.: Towards automatically-tuned neural networks. In: Workshop on Automatic Machine Learning, pp. 58–65 (2016)
100. Klein, A., Christiansen, E., Murphy, K., Hutter, F.: Towards reproducible neural architecture and hyperparameter search (2018)
101. Jin, H., Song, Q., Hu, X.: Efficient neural architecture search with network morphism. CoRR, abs/1806.10282 (2018)
102. Dieleman, S., et al.: Lasagne: first release, August 2015 (2016), 7878. https://doi. org/10.5281/zenodo
103. Zeiler, M.D.: ADADELTA: an adaptive learning rate method. arXiv preprint arXiv:1212.5701 (2012)
104. Kingma, D.P., Ba, J.: Adam: a method for stochastic optimization. arXiv preprint arXiv:1412.6980 (2014)
105. Nesterov, Y.: A method of solving a convex programming problem with convergence rate $o(1/k^2)$ $o(1/k2)$. Sov. Math. Dokl. **27**
106. Duchi, J., Hazan, E., Singer, Y.: Adaptive subgradient methods for online learning and stochastic optimization. J. Mach. Learn. Res. **12**, 2121–2159 (2011)
107. Pham, H., Guan, M.Y., Zoph, B., Le, Q.V., Dean, J.: Efficient neural architecture search via parameter sharing. arXiv preprint arXiv:1802.03268 (2018)
108. Luo, R., Tian, F., Qin, T., Chen, E., Liu, T.-Y.: Neural architecture optimization. In: Advances in Neural Information Processing Systems, pp. 7816–7827 (2018)
109. Boehm, M., et al.: SystemML: declarative machine learning on spark. Proc. VLDB Endowment **9**(13), 1425–1436 (2016)

Travel-Time Computation
Based on GPS Data

Kristian Torp$^{(\boxtimes)}$, Ove Andersen, and Christian Thomsen

Department of Computer Science, Aalborg University, Aalborg, Denmark
torp@cs.aau.dk

Abstract. The volume of GPS data collected from moving vehicles has increased significantly over the last years. We have gone from GPS data being collected every few minutes to data being collected every second. With large quantities of GPS data available it is possible to analyze the traffic on most of the road network without installing road-side equipment.

A very important key performance indicator (KPI) in traffic planning is travel time. For this reason, this paper describes how travel time can be computed from GPS data. Of particular interest is how the travel time is affected by the weather.

The work presented here is an extension of previous work on computing accurate travel time from GPS data. In this paper, the logical data model is explained in more details and the result section showing weather's impact on travel time has been significantly extended with previously unpublished material.

1 Introduction

Estimating travel times in road networks is important for a wide range of applications such as road-network monitoring, driving directions, and traffic planning. When a user requests the travel time from point A to point B, it is expected that the estimated travel time is as accurate as possible particularly for professional drivers such as taxi drivers. However, travel time is complex to estimate because it is affected by many factors such as rush hours, road construction, accidents, and weather.

Even though single trips are sensitive to variations in travel times the impact becomes very significant when it is scaled to the planning of multiple trips for large fleets. Demand Responsive Transport (DRT) is a form of public transport with flexible routes and schedules. A trip is booked in advance to optimize the usage of the vehicles in the fleet. Such trips are being planned from speedmaps, describing the average speeds on a road network during some time interval, e.g., Mondays 10:00 to 12.00. These speedmaps make it is possible to estimate the expected duration of transports to ensure timeliness. Further, with accurate travel-times it is possible to estimate when a vehicle is ready for the next trip.

In DRT, the haulers are often paid by the expected average travel time. The speedmaps are therefore important for calculating the actual cost of a trip and

© Springer Nature Switzerland AG 2020
R.-D. Kutsche and E. Zimányi (Eds.): eBISS 2019, LNBIP 390, pp. 70–92, 2020.
https://doi.org/10.1007/978-3-030-61627-4_4

these maps are the main data foundation for companies like FlexDanmark for optimize the timeliness and the cost of trips.

Until now, the work of determining weather's impact on travel time has mainly focused on analyzing single or a few selected road segments and the data foundation is often fairly limited. In this paper, we determine the weather's impact on a country-size road network using 1.6 billion GPS positions collected from 10,560 vehicles over 5 years. We present a generic model for integrating large scale GPS data with weather information for country-size road networks. We present a model for storing and preparing data for performing a wide variety of travel-time analysis related to weather's impact. The GPS data is map-matched to the road-network of Denmark ($\sim$1.8 million edges). Using this data model, we analyze in detail how the weather conditions *dry*, *fog*, *rain*, and *snow* affects the travel time on the entire road network. The analysis includes determining the correlation between air temperature and travel time and the impact of wind (headwind, tailwind, and crosswind) on travel time. Regional studies are carried out to determine if the weather's impact on travel time is different across regions or different between cities and rural areas.

The content of this paper is an extension of two papers by the authors [5,6]. Compared to [6] the description of the data model is extended with additional details. Compared to [5] the results presented have been significantly extended with previously unpublished material.

The paper is organized as follows. Section 2 describes the data foundation. Section 3 describes the data warehouse star schema used to store the GPS and weather data. Section 4 presents in detail the method used for data integration. A thorough analysis of the weather's impact on travel time is presented in Sect. 5. Section 6 lists related work and Sect. 7 concludes the paper (Fig. 1).

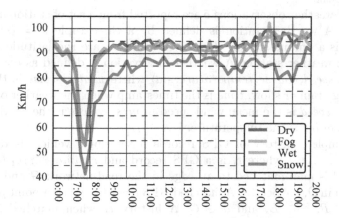

Fig. 1. Average speed by weather

2   Data Foundation

This section describes the GPS data, the map, and the weather data sources integrated to be able to analyze the weather's impact on travel time. First, the data model is presented, next, the concrete data sources used are introduced.

2.1   Data Model

The positions of vehicles are tracked using GPS data. A GPS record, r, is a 6-tuple defined as follows.

$$r = \langle vid, lat, lon, time, speed, course \rangle$$

The tuple contains a unique vehicle id, vid, the position as latitude, lat, and longitude, lon, a timestamp, $time$, a vehicle speed, $speed$, and a compass direction, $course$. The $course$ is used for bidirectional map-matching, i.e., to get a travel times in each direction for road segments that allow two-way traffic. The attributes in r can be collected from most modern GPS devices. The set R denotes all GPS records.

The map foundation is a directed, weighted graph $G = \langle V, E, W \rangle$ where V is a set of vertices and $E \subseteq V \times V$ is a set of edges. Each $v \in V$ is defined by two attributes $v = \langle lat, lon \rangle$ that denote the latitude and longitude. For each edge $e \in E$ we define two attribute $e = \langle course, road\text{-}category \rangle$ where $course$ is the compass direction defined by the straight line between the two vertices that define e (or the endpoints of e). The $road\text{-}category$ is the road category, e.g., motorway or residential road (from the map foundation). The weight $w \in W$ is an array of four values containing the average speed for a segment for four intervals types listed in Table 2 that is free-flow, non-peak, morning peak, and afternoon peak.

A set of weather observations o are reported from a set of stationary weather stations s. A weather station is defined by a three tuple $s = \langle sid, lat, lon \rangle$ where sid is a unique station ID, and lat and lon are the latitude and longitude of the weather station. A weather observation is defined as $o = \langle weather\text{-}class, time, speed, course, temperature, sid \rangle$ where $weather\text{-}class$ is the type of weather, e.g., rain or snow, $time$ is the timestamp when the weather observation is recorded, $speed$ is the mean wind speed in m/s, $course$ is the wind direction, and $temperature$ is the temperature.

The complete history of GPS records for a single vehicle is denoted by $H = [r_1, r_2, \ldots, r_n]$ where r_i is a GPS record and $r_i.time < r_{i+1}.time$. Each GPS record is map-matched to an edge in the road-network G and a weather observation in O, see Sect. 4. A matched GPS record is called a point $p = \langle r, e, o \rangle$ where $r \in R$, $e \in E$, and $o \in O$. A history H when matched is denoted $\hat{H} = [p_1, p_2, \ldots, p_m]$ where $p_i.time < p_{i+1}.time$.

2.2   Data

The map foundation is OpenStreetMap (OSM) [16], from Geofabrik [11]. Road categories associated with the edges are extracted from the OpenStreetMap

Table 1. NOAA weather classes

Weather class	Automated weather report	Manual weather report
Dry	0–5 10 12 18	0–6 10 13–16 18 19
Rain	21–23 40–46 50–53 57 58 60–63 80–84 92 93	20 21 25 50–55 60–65 80–82 91 92
Freezing ppt.	25 47 48 54–56 64–66 74–76 78 89 96	24 27 56–59 66–69 96 97 99
Snow	24 67 68 70–73 77 85–87	22 23 26 70–79 83–90 93 94
Fog	20 30–35	11 12 28 40–49
Thunder	26 90–96	17 29 91–99
Drifting	27–29	7–9 30–39 98
Tornado	99	

Highway tag [17] and the four categories, *motorway, secondary, tertiary,* and *residential* are selected for analysis. The other road categories have not been included in this paper as they show similar results as the four selected categories. In particular, the results for the road category *primary* are very similar to the results for the *secondary* road category. In addition, a map [2] containing the polygons for all cities in Denmark is used for the analysis presented in Sect. 5.

The GPS data coverage for the four selected road categories *motorway, secondary,* and *tertiary* is 99% of road segments are cover. For the *residential* road category, approximately 91% of segments are covered.

Historic weather data is publicly available from the National Oceanic and Atmospheric Administration (NOAA) [3,7]. Denmark is covered by 77 weather stations. This data is used to provide the set of weather stations S and the weather observations O.

The information on wind speed and course is directly available [7]. The weather class associated with each weather observation is retrieved using the approach described next. Each weather observation can have one or more Automatic Weather report (AW) values and Manual Weather report (MW) values [7]. The AW and MW values are integers that identify a specific weather class. We construct eight classes of weather and map the weather codes of AW and MW to these conditions as shown in Table 1. Please note that *Freezing ppt.* (precipitation) includes many descriptions, e.g., ice pellets, hail, freezing rain. Also note that *Drifting* includes both snow and sand. There is no scale on *rain* or *snow* as limited data is available. If both an AW and an MW value exist for a weather observation then MW precedes AW. If more than one AW or MW values exist, the highest value is used.

3   Logical Model

This section describes the logical data model used to store GPS data and the integration with weather information. The section is a significant extension of the description found in [5].

The logical model for the data warehouse is a star schema [13]. The overall design is shown in Fig. 2. The figure shows that there one fact table and 13 dimension tables. The fact table is shown in black and the dimensions tables are grouped and colored by their category.

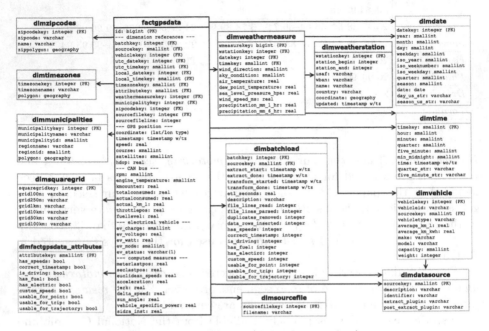

Fig. 2. Logical model (Color figure online)

3.1 Dimensions

The dimensions of the data warehouse are shown in five different colors in Fig. 2. The red dimensions are the spatial dimensions adding locality context to the GPS data. The blue dimensions describe the spatio-temporal weather-related dimensions. The brown dimensions describe date, time, and vehicles. The green dimensions is related to metadata. The gray dimension describes different characteristics and qualities of the data loaded.

The **dimdate** dimension has 12 attributes. It does *not* represent any timezone information. This makes it possible to use one date entry for both local time and UTC time. The primary key is a smart key, i.e., the date 24th of December 2020 has the key value 20201224. The dimension describes typical attributes on dates, along with ISO year, week numbers, and weekdays, where the last week of a year is a full week and a year always starts on a Monday [12]. The dimension has multiple hierarchies since dates both can be grouped into quarters and seasons, e.g., winter and spring. The attributes *year, month,* and *day* describes the date, where the day belongs to the specific month, and the month to the specific year. *weekday* and *day_us_str* describes the days of the

week from Monday through Sunday, where weekday is an integer from 0 to 6 and day_us_str is the name Monday through Friday. *iso_year*, *iso_weeknumber*, and *iso_weekday* stores fiscal years, useful for sorting and aggregating on weeks. *quarter* contains the values 0 to 3, describing the first, second, third, or fourth quarter of the year. *season_us_str* describes Winter, Spring, Summer, or Fall, where *season* stores a corresponding integer for the seasons, from 0 to 3. *date* is a SQL date, without time zones information.

The **dimtime** dimension contains 8 attributes. Like the *dimdate* dimension, it does not contain time-zone information and a timestamp can therefore be used for both local time and UTC time. The primary keys is a smart key and, e.g., the timestamp 0:11 has the smart key 11 and the timestamp 12:34 is the smart key 1234. The attributes are the typical time descriptors, e.g., the attributes *hour* and *minute* describe the current time, *quarter* counts the number of quarters from midnight, i.e. increments by one every fifteen minutes (0 to 95), *five_minute* counts the number of 'five minutes' from midnight, i.e. increments by one every five minutes (0 to 287). *quarter_str* and *five_minute_str* are textually strings for these two groupings of fifteen and five minutes. *min_from_midnight* is a minute counter from midnight (0 to 1439), and time is a SQL time without a timezone. The granularity is one minute. Seconds and sub-seconds are stored as measures on the *factgpsdata* fact table. This is to be able to store GPS data collected with different sampling frequencies, e.g., 0.1, 0.2, 1, 5, or 10 Hz.

Having a finer granularity, e.g., down to seconds, on the *dimtime* dimension is not needed for traffic analysis. Separating date and time dimensions is a common practice [13].

The **dimvehicle** dimension contains a surrogate key and 9 other attributes. A vehicle is identified by *sourcekey* (referencing the *dimdatasource* dimension) and *vehicleid*. The reference to *dimdatasource* is necessary because *vehicleid* only is unique for a certain data source. The other attributes describe general vehicle attributes. namely *make*, *model*, *capacity* (number of passengers), and *weight*. The *average_km_l* and *average_km_kwh* describes average fuel consumption and is computed and updated during ETL for vehicles with fuel or electrical measures.

The **vehicletypekey** references the *dimvehicletype* dimension. The *dimvehicle* dimension is sparsely populated as it is hard to get access to such data from the transportation companies.

The **dimfactgpsdata_attributes** dimension contains 9 Boolean flags computed during the data cleaning fourth each row in the table *factgpsdata*. Each flag describes quality characteristics or use cases of a row of data for data profiling [13] and an attribute can be either true or false, e.g., whether the vehicle is parked, whether timestamp is valid, or whether a GPS observation can be used for map-matching. The flags are *has_speed*, *correct_timestamp*, *is_driving*, *has_fuel*, *has_electric*, and *custom_speed* indicates the quality of data, and the quality is assisting in determining whether a row is used for which of the three map-matching methods, *usable_for_point*, *usable_for_trip*, or *usable_for_trajectory*.

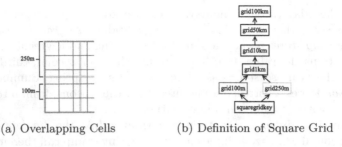

(a) Overlapping Cells   (b) Definition of Square Grid

Fig. 3. Handling grids (Color figure online)

The **dimweathermeasure** dimension is a spatio-temporal dimension that stores weather measurements referenced by the *factgpsdata* table. To save space, *dimweathermeasure* references *dimweatherstation* as in a snowflake schema. The logical schema is therefore not a pure star schema. A weather measure is only valid for a certain time interval where the start is time defined using the *datekey* and *timekey* columns. The current lifetime of a weather measure is one hour, but this can easily be reconfigured. The dimension **dimweathermeasure** references the **dimweatherstation** and the **dimdate** dimensions using *weatherstationkey* and *datekey*, respectively. Each row describes the weather condition at a given station at a given date and hour, *hour*. The weather measures available are *wind_direction*, *sky_condition*, *air_temperature*, *dew_point_temperature*, *sea_level_preasure*, *wind_speed_ms*, *liquid_precipitation_mm_1_hr*, and *liquid_precipitation_mm_6_hr*.

The **dimweatherstation** dimension table stores spatio-temporal information on all weather stations. A weather station has a position and is only valid for a given interval defined by the *station_begin* and *station_end* columns that are date keys referencing the *dimdate* dimension. A weather station is identified by the *usaf* and *wban* attributes that are codes defined by the United States Air Force and the Weather Bureau Army Navy, respectively. *coordinate* is a *geography* data type that stores the (latitude, longitude) of each weather station. This is used to find the nearest station nearest. The primary key is a surrogate key. The *station_begin* and *station_end* describes the dates when a station did start to report weather data and possible when it stopped reporting weather data. *last_updated* stores the date for when information of the weather station was last read from NOAA. *usaf* and *wban* are identifiers for weather stations, where usaf is short for United States Air Force, and wban is short for Weather-Bureau-Army-Navy. *name* and *country* is a name and country for the weather station, *stationid* is an optional identifier for the station. *geog* is a PostGIS geography data type.

The **dimsquaregrid** dimension stores a grid representation of UTM coordinates. The grid is created in multiple layers from the most coarse cell size of 100×100 km, named *100* km, through 50×50 km, named *50* km, 10×10 km, 1×1 km, 250×250 m to the finest cell of 100×100 m, named *100* m. The hierarchy is not strict as the *100* m level is not fully contained by the *250* m level. This can be seen in Fig. 3a, where one *100* m cell (blue) can exist in either 1,

2, or 4 different *250* m cells (green). Therefore the finest level, *100* m, cannot be used as a unique identifier since it can be represented up to 4 times in the dimension table. A *pseudo level* is therefore introduced as shown in Fig. 3b. It has a surrogate key and no further attributes. There is a value for each of the combinations of the *250* m and *100* m levels.

The **dimmunicipalities** dimension is a spatial dimension storing contextual information about municipalities and their spatial regions (polygon). The dimension is pre-loaded and a GPS observation is spatially matched against the polygon. The dimension contains all municipalities in Denmark and is used for matching positions with localities. It consists of 5 elements: A *code* for the municipality, *region_code* for the code of the region, *name* of the municipality, *region_name* for the name of the region, and *geog* storing a PostGIS geography polygon.

The **dimzipcodes** dimension is a spatial dimension for matching GPS observations with zip-code regions. Zip codes are not related to the *dimmunicipality* dimension as one zip code can span several municipalities and one municipality can span multiple zip codes. Zip codes are intervals, and described by *zip_start*, *zip_end*, *zip*, and *name* for the name of the zip region. The polygon is stored in a PostGIS Geography column *geog*.

The **dimtimezones** dimension is a spatial dimension used when GPS observations have UTC timestamps. To determine the corresponding local timestamp, it is necessary to determine in which time zone the GPS observation is in. This is done by finding the nearest timezone polygon. The *tzid* attributes describes the time-zone name and how UTC time is converted to local time.

The **map** dimension is a structure, that is used as a map dimension when a new map is loaded. A map is a graph, storing road segments (edges) with information on intersections (vertices) connecting the road segments. The primary key is a surrogate key and the *segmentid* is an non-unique identifier of the segment (e.g., one OpenStreetMap segment with a single ID might be split into several segments). *startpoint* and *endpoint* describe the connecting vertices of the segment. *meters* describes the segment length, *minutes* describes the travel time one the segment if it represents a ferry route. *segangle* describes the compass direction from the *startpoint* to *endpoint*. *categoryid* is an numeric value describing the category and *category* is a textual description of the category. *speedlimit_forward* and *speedlimit_backward* are the speed limits for the forward and backward directions on the segment. *name* is the road name, *logprime* is the logarithmic value of a unique large prime number, used for when performing trajectory analysis. Finally, *segmentgeo* is a PostGIS Geography type storing the segment as a linestring.

The **dimbatchload** dimension has one entry for each batch load of data. The dimension contains 20 attributes. Five attributes describe the performance of the ETL process and 13 attributes describe the quality of the data. The primary keys is a surrogate key and the *sourcekey* is referencing the *dimdatasource* dimension, because only data from one data source can be loaded at a time. *file_lines_read* has the number of lines read from raw data files, *file_lines_parsed*

has the number of lines that are recognized and parsed by the data extractor. *duplicates_removed* tells how many lines are duplicates (exact vehicle and timestamp already exists in the data warehouse). Duplicates are simply removed. *data_rows_inserted* describes the total number of rows inserted. Some cleaning rules are applied, to classify data during ETL. The number of rows, that meets the different rules of cleaning, is reported to the *has_speeds*, *correct_timestamp*, *is_driving*, *has_fuel*, and *has_electric*. The *custom_speed* describes, how many rows speeds were not available on from data source but where the speed could be computed. This can be done, if high-frequency data is available. *usable_for_point*, *usable_for_trip*, and *usable_for_trajectory* describes how much data is usable for three different map-matching algorithms. *reading_started*, *reading_done*, *cleaning_started*, and *cleaning_done* are timestamps with time zone, describing when data extraction started and ended, and when cleaning and integrating data started and ended. *etl_seconds* describes the total running time in seconds for performing an ETL batch, and *description* is a short description of the data source.

The **dimsourcefile** has a single text attribute that describes the path of the source file a fact originates from. Since the single attribute is not unique for each row of *factgpsdata*, this is located in a dimension [13]. This is useful for verification, debugging, and analysis of data loads. The *dimsourcefile* is not an attribute of *dimbatchload* to allow a batch load to include data from several source files.

The **dimdatasource** dimension describes the data sources available as several organizations provide data. Each data source has a different data format (column in a CSV or XML file) and data is handled fairly differently. The *identifier* is a textual identifier for a data source, *etl_plugin* describes which ETL plugin is used, and *etl_postprocess_plugin* is an optional plugin, that can be executed after data is loaded. The *description* is a short description of the data source.

3.2   Fact Table

The **factgpsdata** fact table contains one row for each GPS observation. All GPS data is represented here. This means that even though some data is classified as being invalid, e.g., due to a timestamp in the future, it is still being stored in the data warehouse for analytical purposes. Note that due to the metadata stored, it is easy to include/exclude such data when doing analysis.

The fact table references all the dimensions, see Fig. 2. The table has the following columns: First, there is a unique *id*. The *id* is included to make it possible to reference a fact from another fact table (we do this in other work outside the scope of this paper). The *id* is followed by the dimension references. *utc_datekey* and *local_datekey* reference the *dimdate* dimension to represent the UTC date and the local date, respectively. Similarly, *utc_timekey* and *local_timekey* stores the UTC time and the local time, respectively.

The *weathermeasurekey* is added during ETL transformation but it may happen that the external weather entry is not yet available. Therefore this fact table

can contain early arriving facts [13] which are updated later when the weather entry becomes available. The source data is identified by a dimension reference in *sourcefilekey* and a degenerate dimension [13] in *sourcefileline*.

After the dimension references, the GPS observation data is represented. The *coordinate* is a latitude/longitude, which is a fixed-size data type that stores latitude and longitude coordinates. This is useful because it enables us to use a range of spatial functions for analysis. The *seconds* is an addition to the *utc_timekey* and *local_timekey*, where the seconds and milliseconds of a timestamp is stored as a real value.

The complete timestamp is stored in *timestamp*, which is useful for sorting data and to calculate the vehicle's (average) speed between two GPS observations. Eight measures from the CAN bus system can be represented. The four measures (*"ev_"*) are values only available from electric vehicles.

4   Data Cleaning Method

In this section, we describe how data is prepared to produce the results presented in Sect. 5. The data foundation presented in Sect. 2 is referenced intensively in this section. We use the dot-notation to access a specific attribute in a tuple, e.g., $r_i.lat$ denotes the latitude of a GPS record r_i.

The cleansing method consists of three steps. First, a GPS record is map-matched to the road network. Second, the record is matched to a weather observation. Finally, from the matched points P and the weight W associated with each edge in the graph G are computed. We describe these three steps in the following. Again part of the description is from previous work.

4.1   Map-Matching

Informally, a GPS record $r \in H$ is map-matched to the nearest edge $e_{mm} \in E$. The GPS records of H are low-frequent with a sampling rate of 15–60 s between each record, thus a trajectory cannot be reconstructed from a set of GPS records. A record is map-matched with the conditions that 1) the distance between e_{mm} and r is maximum 25 m and 2) that the angle difference between e_{mm} and r is up to 22.5°. However, because GPS devices generally have problems computing the compass directions at low speed there is the additional condition that if the speed is below 2 km/h then the edge that the previous position record in H was map-matched to is reused for the current record.

The values 25 m, 22.5°, and 2 km/h have been found by analyzing one month of GPS data. Figure 4a shows how many GPS records can be map-matched when limiting the allowed distance between e_{mm} and r. It can be seen, that allowing a distance of up to 5 m will accept 77% of data to be map-matched, while if allowing up to 25 m will accept 97% of data to be map-matched.

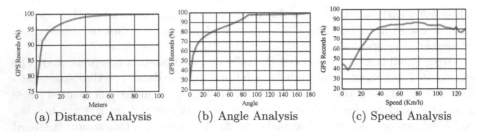

(a) Distance Analysis   (b) Angle Analysis   (c) Speed Analysis

Fig. 4. Analysis of map-matching parameters

Figure 4b shows how many GPS observations are accepted when limiting the angle offset between a e_{mm} and r, searching within a 25 m radius. By allowing an angle offset of up to 22.5° a total of 76% of GPS measurements can be matched to a road segment.

Using an offset angle of 22.5° Fig. 4c shows how many GPS records are accepted by this angle at different speeds. It can be seen that the course of GPS positions have some imprecision at low speeds. A threshold of 2 km/h has been chosen when using the GPS course.

The algorithm for map-matching the GPS records r in a history H to the road network G is described in Algorithm 1. The input to the map-matching algorithm mm is a history H of GPS records and a road network (a map) G. An empty matched history $\hat{h}$ is created. Then the records r_i in the history H are mapped in sequential order. For each record r_i the function *nearest_neighbor* finds the nearest edge e_{mm} in G. However, the distance between the GPS record and e_{mm} must be less than 25 m and the angle between the $r_i.course$ and $e_{mm}.course$ must be less than 22.5°. If these two conditions are not fulfilled it is checked if the speed is below 2 km/h. If yes, then the edge of the previous point is reused for p_i. Otherwise, the GPS record r_i is discarded and is not a part of $\hat{h}$.

We denote the distance between a position record, $p \in P$, and a road segment, $e \in E$, as $dist(p, e)$. Concretely, we use the Euclidean distance in the implementation.

The angular difference *angle* between two position records, where $p \in P$, $e \in E$ is defined as follows.

$$angle(p, e) = \begin{cases} |p.course - e.course| & \text{if } |p.course - e.course| \leq 180 \\ 360 - |p.course - e.course| & \text{otherwise} \end{cases}$$

4.2   Weather Class

To determine the weather's impact on travel time we also match each GPS record to the weather class at the nearest weather station at the time the GPS record was recorded. The work presented here is a generalization and an extension of existing work [5].

Algorithm 1. Map-Matching Algorithm

function MM(H,G) *//H=GPS record history, G=map*
  $\hat{h} \leftarrow [\,]$
  for all $r_i \in H$ **do**
    $p_i.r \leftarrow r_i$ *//p_i stores map-matched information for r_i*
    $e_{mm} \leftarrow nearest\_neighbor((r_i.lat, r_i.lon), G)$
    if $dist(r_i, e_{mm}) \wedge angle(r_i.course - e_{mm}.course) < 22.5°$ **then**
      $p_i.e \leftarrow e_{mm}$
    else if $r_i.speed < 2\text{km/h} \wedge p_{i-1}.e = e_{mm}$ **then**
      $p_i.e \leftarrow e_{mm}$
    else
      continue *//Skip this record*
    end if
    $\hat{h}.append(p_i)$
  end for
  return $\hat{h}$
end function

The algorithm for matching a position to a weather observation is listed in Algorithm 2. It takes a map-matched history $\hat{H}$, produced by Algorithm 1, and the set of weather stations S as input. For each point p_i the nearest weather station is found. If this weather station is more than 200 km away the point is removed from $\hat{H}$. Otherwise, the weather observation, o_{mm}, valid at the nearest station, s_{mm}, is found and assigned to the observation attribute of the p_i point. Note that Algorithm 2 is a procedure that changes the parameter $\hat{H}$ and does not have an explicit return value.

To study the effects of the wind, we define three wind attack classes, that is tailwind, crosswind, and headwind. The three classes are defined by an angle β describing the accepted offset from direct tailwind, crosswind, or headwind.

Figure 5 illustrates the angles, β_t for tailwind, β_c for crosswind, and β_h for headwind. The yellow arrow illustrates a vehicle and its driving direction. When the angle between the driving direction and the wind is less than β_c the wind is classified as tailwind. When the angle between the opposite direction of the driving direction and the wind direction is less than β_h then wind is classified as headwind. When the angle between the perpendicular direction of the driving direction and wind direction is less than β_c then the wind is classified as crosswind. The mean wind speed is classified into four speed groups. These groups are 1–5 m/s, 6–10 m/s, 11–15 m/s, and 16- m/s. These groups describe calm, light, moderate, and heavy wind conditions.

4.3 Speedmaps

The results presented in Sect. 5 are based on speedmaps, which are the weighted, directed graph presented in Sect. 2. In this section, we describe how the weights W are added to a graph G.

Vehicles that move slowly will produce more GPS records on an edge (road segment) than vehicles that move fast. To avoid the slow-moving vehicles

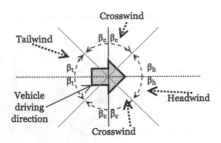

Fig. 5. Wind direction identifiers

Table 2. Time-of-day intervals

Interval	Name
00:00–06:00	free-flow
06:00–07:30	non peak
07:30–08:15	morning peak
08:15–15:00	non peak
15:00–16:30	afternoon peak
16:30–20:00	non peak
20:00–24:00	free-flow

Algorithm 2. Weather-Matching Algorithm

---

procedure WEATHER-MATCH($\hat{H}, S$) //$\hat{H}$=Map-matched history, S=weather stations
  for all $p_i \in \hat{H}$ **do**
    $s_{mm} \leftarrow nearest\_station((p.r.lat, p.r.lon), S)$
    if $dist(s_{mm}, p.r) > 200$km **then**
      $\hat{H}.remove(p_i)$
      continue //*Weather station too far away*
    end if
    $o_{mm} \leftarrow weather\_at\_station(s_{mm}, p.r.time)$
    $\hat{h}.o \leftarrow o_{mm}$
  end for
end procedure

---

weighing too much, the GPS records for a single vehicle that traverses an edge is grouped and the average speed for the GPS records is computed. The idea is shown in Fig. 6. Here the triangles illustrate GPS records from a slowly moving record and the circle illustrates a single GPS record from a fast-moving vehicle. The speed of the fast-moving vehicles is directly associated with the edge e_1. For the slow-moving vehicles, only the average speed of the GPS records are associated with the edge. We showed earlier, that more than 90% of the map is covered by GPS data. To cover the remaining segments, with no GPS coverage, we have implemented a trivial spatio-temporal smearing algorithm, where speeds from similar road segments are utilized. Further description is omitted due to space limitations.

We only use GPS records from workdays, i.e., Monday through Friday, and due to congestion, we split the time-of-day into the non-overlapping time intervals shown in Table 2. These time intervals are developed in association with road authorities and can easily be changed.

The time-intervals are used to create an average speed within an interval, i.e., four intervals in total for each edge. Thus for each edge, the weight W is an array with four values, which are the average speed in the intervals, free-flow, non peak, morning peak, and afternoon peak.

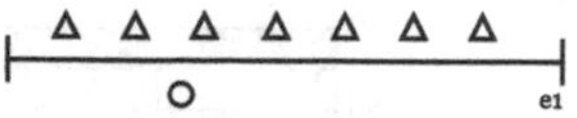

Fig. 6. Weighted average for speed

The speed for the time intervals is used to study the relation between congestion and weather class. To do this we introduce a congestion level, C, shown by Eq. 1.

$$C = \frac{v_{free\text{-}flow} - v_{morning\_peak}}{v_{free\text{-}flow}} \tag{1}$$

C is zero there is no congestion on an edge and when C is 1 the traffic has come to a complete stop. The congestion level is calculated from the relative speed difference between free-flow speed, $v_{free\text{-}flow}$, and morning peak speed, $v_{morning\_peak}$.

5   Results

The results related to weather are presented in a top-down manner. We first analyze the weather implications on the entire road network, next the focus is to compare weather impact in urban versus rural areas. Finally, the weather impact on individual streets is examined.

Next, the impact of congestion and weather is studied. All trips with weather information from 2014 are analyzed with respect to different weather conditions. Wind effects on the traffic are studied to determine the impact of the wind attack angle and then selected road segments are analyzed to determine the impact of tail-, cross-, and headwind.

5.1   Weather Classes to Include

Denmark is a country with limited climatic differences between regions. The climate is temperate coastal climate, i.e., often mild winters, approximately 120 days of precipitation annually, and rarely dangerous weather, e.g., hurricanes and tornadoes. Figure 7 shows the distribution of GPS reports from vehicles within eight weather classes. Denmark is dominated by *dry* weather and *rain*, along with some periods of *fog* and *snow*. Due to this uneven distribution only the top four classes, *dry, fog, rain*, and *snow*, are used for further analysis. *Freezing ppt.*, *drifting, thunder*, and *tornado* are removed from the rest of the weather analysis as they rarely occur.

From Fig. 8 it can be seen that *snow* is typically present from November through March, while *fog* is fairly evenly distributed over the year. *Dry* and *rain* are varying across the seasons from year to year. February, March, October, and November are the driest months with more than 50% of observations being *Dry*. Most of the months the observations are fairly evenly distributed between being *Dry* and *Wet* or *Snow* having August and December with less than 40% *Dry* weather. *Fog* is present in 5–10% of the observations and *Snow* observations are mainly present from November through March.

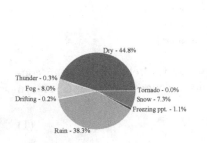

Fig. 7. Weather distribution

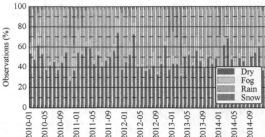

Fig. 8. Monthly weather distribution

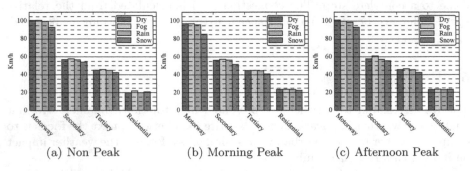

(a) Non Peak   (b) Morning Peak   (c) Afternoon Peak

Fig. 9. Average speed on road categories in different intervals

Road Categories and Weather. The average speeds on all roads in an entire road network is a good indicator of the weather's impact in general.

Figure 9a shows the average speed on the four road categories in non-peak intervals, depending on different weather classes. It can be seen that *dry*, *fog*, and *rain* are very comparable and the speed only varies 2% on all road categories. Snow has an impact of up to 8% on *motorway*, *secondary*, and *tertiary*. On *residential* roads, there is no measurable impact of *snow*. The effects are similar when looking at the morning peak, Fig. 9b, and the afternoon peak, Fig. 9c, where *dry*, *fog*, and *rain* are comparable, while *snow* leads to increased lower speeds in morning traffic.

Road Categories and Temperature. The relation between temperature and average speed on the road categories are relevant to examine to determine if they are related. Figure 10 shows the distribution of weather observations depending on the temperature. This figure shows that *snow* is most likely to occur between 0 and −6 °C. When temperatures drop below −6 °C the number of observations with precipitation decreases while fog increases below −6 °C.

Figure 11 shows the average speeds on the four road categories at different temperatures, ranging from −15 to +25 °C. It can be seen that from +3 to +25 °C there is little or no variation while below 3 °C there is a decline in the speed of 5–9% on all road categories. The drop in speed is most likely due

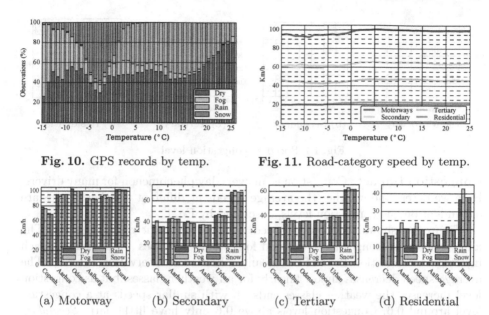

Fig. 10. GPS records by temp.   **Fig. 11.** Road-category speed by temp.

(a) Motorway   (b) Secondary   (c) Tertiary   (d) Residential

Fig. 12. Regional average speeds

to an increased risk of frosty and icy roads. Also, sleet and snow occur when temperatures get close to or less than $0\,°C$.

In general, it can be seen that speed drops when the temperature gets close to $0\,°C$, but when temperatures drop below $-10\,°C$ the speed on *motorway* and *tertiary* increases by 2–3 km/h. This can be due to snow and sleet mainly occurring between -6 and $+2\,°C$, which can be seen in Fig. 10.

Regional Impact. We study the impact of weather in the four largest cities in Denmark and compare these to all urban and all rural areas. Morning peak interval will be the outset for analyzing the changes in speed.

Figure 12a shows that the *motorway* speed varies significantly across the regions. In Copenhagen and Odense *dry* weather leads to increased speeds, while for the rest of the regions the weather impact is very limited. *Secondary* roads, Fig. 12b, have different speeds over the different regions, while weather impact is very limited, except for *fog*, which shows a tendency to slightly faster speeds. The speed on *Tertiary* and *residential* roads, Fig. 12c and Fig. 12d, is limited affected by the weather.

There are regional differences in speeds, which makes it clear that road category comparisons should be done carefully. There is only a little difference between *dry* weather and *rain*. *Fog* tends to, unexpectedly, show increased speeds, which can be due to *fog* can be a very local phenomenon and weather stations covers a relatively large area.

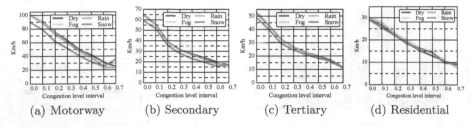

(a) Motorway   (b) Secondary   (c) Tertiary   (d) Residential

Fig. 13. Speed at congestion level

Congestion Impact. Congestion is an everyday phenomenon for many drivers. Here we determine whether the average speeds at different weather classes are dependent on the congestion levels, described by Eq. 1.

Figure 13a shows the average speed in morning traffic on motorways. The x-axis shows the congestion levels and four weather classes are present for each interval. A congestion level of 0.1 means a half-open interval of [0.1,0.2). The average speed decreases almost linearly for all weather classes, but as congestion level increases the weather classes ends up with similar speeds at a congestion level around 0.6. Congestion levels above 0.6 only have little data. *Secondary* and *tertiary* roads, Fig. 13b and Fig. 13c, show similar results where the gap between the average speeds closes in as congestion levels increase. *Residential* roads, Fig. 13d, indicates that there is limited difference between the four weather classes.

In general, it can be seen from the results that as congestion increases the impact on different weather classes decreases. Congestion can though be seen as a primary factor here and weather impact becomes a minor factor.

Speed Distribution. The average speed on a road segment is an indication of how fast the average vehicle is driving but does not tell anything about the distribution of the speeds. Four road segments have been selected for detailed analysis in Fig. 14, two motorway segments, a rural segment, and an urban segment. All four segments are analyzed using data from Monday through Friday between 9 a.m. and 14 p.m.

For each boxplot, a blue box shows the quartiles and a red line the median. The red square shows the mean and the whiskers and fliers show the reach past the first and third quartile of 1.5 (upper whisker is $Q3 + 1.5 * (Q3 - Q1)$). From Fig. 14a through Fig. 14d it can be seen that for all four road segments there is little to no variation between the speed distribution between *dry, rain,* and *fog*. Speeds at *snow* are in general a bit slower but the quartiles show no increased variation in the data, except for outliers with high speeds when *snow*.

Road Stretch Analysis. While aggregated analysis is good at providing an overall view of the weather's impact, more detailed analysis can give a deeper insight on individual road. We study the weather's impact on four different

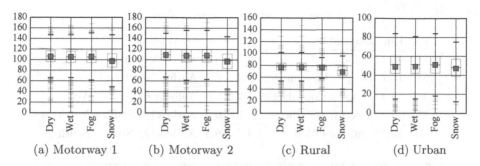

(a) Motorway 1   (b) Motorway 2   (c) Rural   (d) Urban

Fig. 14. Vehicle speed statistics (Color figure online)

Table 3. Travel minutes on weather conditions

Type	Trips	Weather speedmap, min.	Baseline Speedmap, min.	Relative gain
Dry	434,027	8,205,992	8,295,145	−1.07%
Fog	77,265	1,512,784	1,536,979	−1.57%
Rain	418,827	7,794,186	7,848,050	−0.69%
Snow	50,319	985,827	919,851	7.17%
Total	997,578	18,498,789	18,600,025	−0.54%

motorway stretches, labeled $M1$ through $M4$, two rural stretches, labeled $R1$ to $R2$ and four urban stretches, labeled $U1$ through $U4$.

Figure 15 shows heat maps of the routes in the morning traffic, where D is Dry, F is Fog, R is Rain, and S is Snow. Dry speeds are the baseline speeds and the cells are colored by their relative difference to dry speed, that is the percentage for each weather class. Yellow/red indicates slower speeds and dark green/blue indicates faster speeds than dry weather. Figure 15a shows only limited impact by *rain* for all routes. *Fog* shows a significant impact on three routes, $M2$, $M3$, and $U1$, while only limited impact for the remaining seven routes. Only *snow* causes a significant reduction in speed by up to 13.8% for the $M1$, $M3$, $M4$, $R1$, and $R2$. The urban roads are only affected by *snow* in a less degree. Morning peak speeds, Fig. 15b, shows that *fog* is quite often faster than *dry* weather. Speeds while *snow* are more impacted and reduced by up to 27%. Afternoon peak speeds, Fig. 15c, also shows tendencies to faster speed at *fog* similar to non-peak, with relative speed differences of up to 16%.

When comparing road stretches it is interesting that afternoon peak is more closely related to non-peak intervals than morning peak. This is probably due to that morning peak traffic is more condensed where the afternoon peak traffic is stretched over a longer interval.

Trip Analysis. A set of trips are examined to study how the travel time is affected by the weather conditions. The travel time in *dry* weather conditions is used as the baseline (Table 4).

Table 3 shows that the trips are divided into mainly *dry* and *rain*. The weather speedmap is 0.7 to 1.6% faster than the baseline speedmap for trips

Table 4. Road stretches to analyze

ID	Km	Description	ID	Km	Description
M1	98.4	Aalborg - Aarhus	R2	13.6	Mou - Aalborg
M2	16.1	Koege - Copenhagen	U1	5.7	Aarhus North - C
M3	13.4	Farum - Copenhagen	U2	2.6	Aalborg O2 East - West
M4	19.9	Tylstrup - Aalborg	U3	9.4	Aarhus O2 South - North
R1	13.1	Aalborg - Mou	U4	9.5	Aarhus O2 North - South

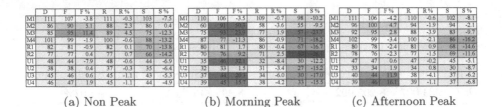

(a) Non Peak   (b) Morning Peak   (c) Afternoon Peak

Fig. 15. Weather impact on road stretches (Color figure online)

during *dry*, *fog*, and *rain* conditions. During *snow* the weather-dependent speedmap yields 7.2% longer travel time compared to the baseline speedmap. When summing up all trips it shows that if weather-dependent speedmaps are used, compared to a non-weather speedmap, a total reduction in travel time of 0.5% can be expected when planning trips, that is a saving of approximately 101 K travel minutes.

5.2 Wind Analysis

This section studies the impact of the wind. The impact will be measured by the average speed on different conditions. The wind is classified into four groups, as shown by Table 5, where the number of GPS records for every wind-speed class can be seen. The wind speeds are mean wind and gusts of stronger speed are likely to occur.

Effect of Wind. To analyze the impact of wind we will study the wind impact on *motorway* segments as vehicles tend to have relatively stable speeds on these segments.

Figure 16a shows the effect of tailwind on motorway stretches. The figure shows that the speed is slightly affected by the angle of the wind used. A narrower angle means only a very direct tailwind is accepted, while a broader angle means accepting more crosswind. Accepting a wider angle only yields a decreased speed of 1 km/h, except for very strong winds where speeds decrease by 2 km/h going from a β of 10 to 90. Figure 16b shows the impact of increasing the angle for accepting crosswinds. It can be seen that for winds of 11–15 and 16- m/s there is an impact when increasing β, thus accepting evenly more tail- and headwind. It can be seen that a wider angle yields faster speeds for 16- m/s winds, which

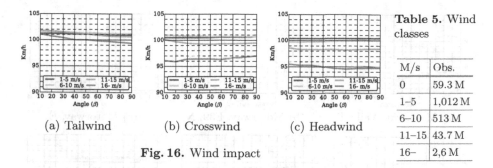

M/s	Obs.
0	59.3 M
1–5	1,012 M
6–10	513 M
11–15	43.7 M
16–	2,6 M

(a) Tailwind   (b) Crosswind   (c) Headwind

Table 5. Wind classes

Fig. 16. Wind impact

indicates tailwind has a stronger effect than headwind. Figure 16c shows accepting more crosswind has a little impact in speed, mainly at 16- m/s, though speed is only varying 1 km/h.

Due to the analysis of the wind attack angle, we decide on an angle (β) of 45°. That yields a total angle of 90° head and tailwind while crosswind covers 90 degrees to the left of the vehicle and 90° to the right of the vehicle, thus 180°. Comparing Fig. 16a through Fig. 16c it can be seen, that for wind speeds ≤ 10 m/s there is no significant difference between tail-, cross-, and headwind, while stronger winds of 11–15 and 16- m/s indicate faster speeds of tailwind than crosswind and faster speeds of crosswind than headwind.

Vehicle Type Impact. Two road segments have been selected for performing detail analysis of the wind impact, along with an aggregated analysis on all *motorway*segments, Fig. 17. The driving direction is indicated in the caption, N for North, E for East, and so on.

In general, it can be seen, that vehicle speed decreases at crosswind and headwind when the wind speed increases. Figure 17e shows though that this road stretch is not impacted by wind, except at wind speeds of 11 m/s or stronger. Tailwind show tendencies to often result in slightly increased speeds. Figure 17a indicates that cars traveling at faster speeds than minibuses and wind speed only have limited influence. The same tendency shows at Fig. 17b through Fig. 17e, where cars and minibuses show comparable speeds while they only show little impact by the wind speed. The two motorway bridges at Fig. 17c and Fig. 17d indicates that there are slightly faster speeds when tailwind is available and as cross and headwind increases the speed is reduced. The primary road, Fig. 17e shows that minibuses are driving faster than cars at this stretch, though when wind increases in strengths the headwind of minibuses have an impact and speed is reduced by 5 to 10 km/h at 11 and 12 m/s. Finally, Fig. 17f shows all motorway stretches combined. Here there is a clear tendency to that when wind increases in strength, from 5 m/s and up, the speed decreases slightly. This is the case for both cars and minibuses for both head, cross, and at some extend tailwind, though do tailwind seems to be less impacted.

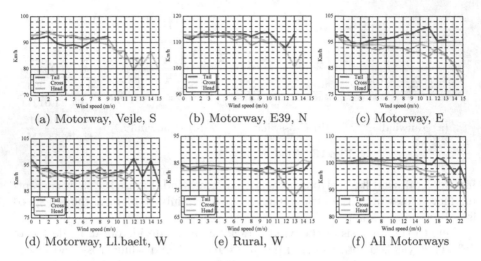

(a) Motorway, Vejle, S   (b) Motorway, E39, N   (c) Motorway, E

(d) Motorway, Ll.baelt, W   (e) Rural, W   (f) All Motorways

Fig. 17. Speed and wind relation

6   Related Work

The field of analyzing weather impact on vehicle speeds have been studied for years. Most existing work, [1, 4, 8–10, 14, 15, 18–21, 23], utilizes induction loop detectors to obtain traffic data. The works study the impact of weather on travel time, traffic flow, and traffic levels. In general, they find that rain has a varying impact on travel time while snow can have a great impact. As the studies are limited to induction loop detectors, the studies are mainly on single or few road segments. Most of the work utilizes data for shorter periods, weeks or months, while some have data for multiple years.

In contradiction to using loop detectors, [22] uses an Automated Number Plate Recognition (ANPR) system for obtaining similar results for London, showing that temperatures below 0 °C imply delays and the intensity of rain and snow can impact speeds.

GPS data has been utilized by [25] where 8,000 taxis provide 800,000 records over 4 months in Hongzhou, China. They propose a prediction framework and while doing so they analyze weather impact. Another work, [24], utilizes 10 M GPS records over 2 months from 1,570 taxis in Nagoya City, Japan.

Existing work for analyzing weather impact on road networks often suffer from at least one of two factors. Firstly, most related work only utilizes data for shorter periods, e.g., a few months, making the analysis suffering from seasonal variations. Secondly, existing studies only perform analysis on reduced samples of a road networks, either due to fixed measuring stations (loop detectors, ANPR detectors, etc.) or spatially limited extent of GPS data.

7   Conclusion

This paper presents a large-scale nation-wide study of how weather impacts the speed in road networks. 1.6 billion GPS data is collected from 10,560 vehicles over five years from 2010 through 2014 across all of Denmark. The data is integrated with OpenStreetMap and detailed weather information from NOAA.

A generic data model is presented which has a global scope and is applicable if a set of GPS data and a road network graph is present.

Using the weather classes *dry, fog, rain,* and *snow* we show that *snow* has the greatest impact, primarily on *motorway, secondary,* and *tertiary* roads with a reduction in speed of up to 27%. *Residential* roads show only little to no impact on *snow.* For the other weather classes (*dry, fog,* and *rain*) there are only smaller differences across all road categories.

Analyzing regional differences we show that the weather impact is different between cities. Congestion and *snow* both affect speed negatively. However, at higher congestion levels there is little difference between *snow* and the other weather classes. The outside temperature is correlated with speeds, as low temperature implies up to 9% reduction in speed. Similarly, we show that wind can reduce speeds with up to 19%.

In conclusion, to compute or predict the average speed accurately it is necessary to take into consideration, the three factors, weather class, temperature, and wind speed.

Acknowledgment. This paper is partly based on papers coauthered with Benjamin B. Krogh and Christian S. Jensen.

References

1. Agarwal, M., Maze, T.H., Souleyrette, R.: Impacts of weather on urban freeway traffic flow characteristics and facility capacity. Technical report (2005)
2. Agency, D.G.: Kortforsyningen. http://download.kortforsyningen.dk/
3. Agency, N.: NOAA - north oceanic and atmospheric administration. http://www.noaa.gov/
4. Akin, D., Sisiopiku, V.P., Skabardonis, A.: Impacts of weather on traffic flow characteristics of urban freeways in Istanbul. Procedia Soc. Behav. Sci. **16**, 89–99 (2011)
5. Andersen, O., Krogh, B.B., Thomsen, C., Torp, K.: An advanced data warehouse for integrating large sets of GPS data. In: Proceedings of the 17th International Workshop on Data Warehousing and OLAP, DOLAP 2014, Shanghai, China, 3–7 November 2014, pp. 13–22 (2014). https://doi.org/10.1145/2666158.2666172
6. Andersen, O., Torp, K.: A data model for determining weather's impact on travel time. In: Database and Expert Systems Applications - 27th International Conference, DEXA 2016, Porto, Portugal, 5–8 September 2016, Proceedings, Part II, pp. 437–444 (2016). https://doi.org/10.1007/978-3-319-44406-2_37
7. Center, N.C.D.: Federal climate complex data documentation for integrated surface data. Technical report (2015). ftp://ftp.ncdc.noaa.gov/pub/data/noaa/ish-format-document.pdf

8. Chung, E., Ohtani, O., Warita, H., Kuwahara, M., Morita, H.: Does weather affect highway capacity. In: 5th International Symposium on Highway Capacity and Quality of Service, Yakoma, Japan (2006)
9. Datla, S., Sahu, P., Roh, H.J., Sharma, S.: A comprehensive analysis of the association of highway traffic with winter weather conditions. Procedia Soc. Behav. Sci. **104**, 497–506 (2013)
10. Edwards, J.B.: Speed adjustment of motorway commuter traffic to inclement weather. Transp. Res. Part F Traffic Psychol. Behav. **2**(1), 1–14 (1999)
11. GeoFabrik: Openstreetmap data extracts. http://download.geofabrik.de/
12. ISO: ISO 8601:2004 data elements and interchange formats. Information interchange. Representation of dates and times. ISO, pub-ISO (2005)
13. Kimball, R., Ross, M.: The Data Warehouse Toolkit: The Definitive Guide to Dimensional Modeling. Wiley, Hoboken (2013). http://books.google.dk/books?id=4rFXzk8wAB8C
14. Mashros, N., Ben-Edigbe, J., Hassan, S.A., Hassan, N.A., Yunus, N.Z.M.: Impact of rainfall condition on traffic flow and speed: a case study in johor and terengganu. Jurnal Teknologi **70**(4), 1–5 (2014)
15. Maze, T., Agarwai, M., Burchett, G.: Whether weather matters to traffic demand, traffic safety, and traffic operations and flow. Transp. Res. Rec. J. Transp. Res. Board **1948**, 170–176 (2006)
16. OpenStreetMap. http://www.openstreetmap.org
17. OpenStreetMap: Key: highway - openstreetmap. http://wiki.openstreetmap.org/wiki/Key:highway
18. Rakha, H., Farzaneh, M., Arafeh, M., Hranac, R., Sterzin, E., Krechmer, D.: Empirical studies on traffic flow in inclement weather. Virginia Tech Transportation Institute (2007)
19. Saberi, K.M., Bertini, R.L.: Empirical analysis of the effects of rain on measured freeway traffic parameters (2010)
20. Smith, B.L., Byrne, K.G., Copperman, R.B., Hennessy, S.M., Goodall, N.J.: An investigation into the impact of rainfall on freeway traffic flow (2004)
21. Thakuriah, P., Tilahun, N.: Incorporating weather information into real-time speed estimates: comparison of alternative models. J. Transp. Eng. **139**(4), 379–389 (2013). https://doi.org/10.1061/(ASCE)TE.1943-5436.0000506
22. Tsapakis, I., Cheng, T., Bolbol, A.: Impact of weather conditions on macroscopic urban travel times. J. Transp. Geogr. **28**, 204–211 (2013)
23. Tu, H., van Lint, H.W., van Zuylen, H.J.: Impact of adverse weather on travel time variability of freeway corridors. In: Transportation Research Board 86th Annual Meeting (2007)
24. Wang, L., Yamamoto, T., Miwa, T., Morikawa, T.: An analysis of effects of rainfall on travel speed at signalized surface road network based on probe vehicle data. In: Proceedings of the Conference on Traffic and Transportation Studies, ICTTS, Xi'an, China, pp. 2–4 (2006)
25. Zhang, R., Shu, Y., Yang, Z., Cheng, P., Chen, J.: Hybrid traffic speed modeling and prediction using real-world data. In: 2015 IEEE International Congress on Big Data (BigData Congress), pp. 230–237 (2015)

Laplacian Matrix for Dimensionality Reduction and Clustering

Laurenz Wiskott[✉][iD] and Fabian Schönfeld

Institut für Neuroinformatik, Ruhr-Universität Bochum, Bochum, Germany
laurenz.wiskott@rub.de, fabian.schoenfeld@ini.rub.de
https://www.ini.rub.de/

Abstract. Many problems in machine learning can be expressed by means of a graph with nodes representing training samples and edges representing the relationship between samples in terms of similarity, temporal proximity, or label information. Graphs can in turn be represented by matrices. A special example is the Laplacian matrix, which allows us to assign each node a value that varies only little between strongly connected nodes and more between distant nodes. Such an assignment can be used to extract a useful feature representation, find a good embedding of data in a low dimensional space, or perform clustering on the original samples. In these lecture notes we first introduce the Laplacian matrix and then present a small number of algorithms designed around it for data visualization and feature extraction.

Keywords: Dimensionality reduction · Embedding · Clustering · Spectral graph theory · Laplacian matrix · Laplacian eigenmaps (LEM) · Locality preserving projections (LPP) · Spectral clustering

1 Intuition

The Laplacian matrix can be used to model heat diffusion in a graph. Its theory can thus be understood intuitively with the help of the heat diffusion analogy.[1]

1.1 Heat Diffusion Analogy of Laplacian Eigenmaps

First consider a very simple heat diffusion analogy for nonlinear dimensionality reduction from 2D to 1D with the Laplacian eigenmaps algorithm. Fig. 1 (left) shows seven points in 2D, labeled A through G. Their position might not be very meaningful but we assume that we have some similarity function that induces

---

L. Wiskott—This contribution is a modified version of [13].

[1] This section is meant to give an intuitive introduction into the Laplacian matrix, Laplacian eigenmaps, and spectral clustering. It is not necessary to understand the remainder of the lecture notes but hopefully makes it easier. If you are short on time and rich in math and machine learning background, you might prefer to skip it.

© Springer Nature Switzerland AG 2020
R.-D. Kutsche and E. Zimányi (Eds.): eBISS 2019, LNBIP 390, pp. 93–119, 2020.
https://doi.org/10.1007/978-3-030-61627-4_5

relationships between these points. This results in a simple undirected graph with seven nodes and six edges in this example. We see already that the graph is a simple linear graph, a chain, but in high dimensions with many more nodes and a slightly more complicated structure, this might not be so obvious anymore.

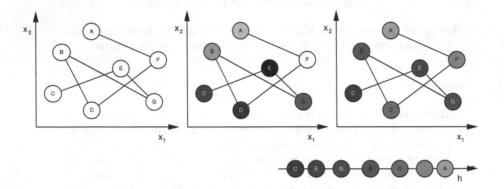

Fig. 1. Heat diffusion analogy of the Laplacian eigenmaps algorithm.

The heat diffusion analogy now says that nodes are considered heat reservoirs and heat can diffuse from one node to neighboring nodes via the edges, but no heat gets lost or added. So, let us randomly initialize the nodes with arbitrary temperatures, Fig. 1 (middle). What happens if we wait? Well, it is obvious that heat diffuses from warmer to colder nodes until temperature has balanced out completely. It is also obvious that local temperature differences balance out quickly, while global temperature differences between distant nodes (distant in terms of the graph connectivity) take more time to balance out. So if one measures the temperatures quite late in the process, one finds a distribution like the one shown in Fig. 1 (right). One end of the chain is slightly warmer than the other end, and from one end to the other there is a monotonic decrease of temperature. This is interesting, because if one now plots the seven points again, but now in a 1D space according to their temperature, one gets the plot in Fig. 1 (bottom right). The points are nicely ordered by their position in the linear graph. This is much better for visualization and interpretation and possibly further processing of the points, since the position in space now reflects similarity relations well. (The details of the spacing reveal a flattening of the temperature profile towards the ends, an effect that takes more effort to understand intuitively and is beyond the scope of this introduction.)

This is essentially how the Laplacian eigenmaps algorithm works, except that one does not really use heat diffusion but finds the resulting heat distribution analytically in a more efficient and robust way. It is also possible to map the points into a 2D or even higher-dimensional space by taking more than one heat diffusion mode into account.

1.2   Heat Diffusion Analogy of Spectral Clustering

For a heat diffusion analogy of spectral clustering consider a different connectivity of the graph, like the one shown in Fig. 2 (left). The difference to the example above is that now the graph has two disconnected subgraphs. No heat can diffuse from one subgraph to the other. If one waits long enough, the temperature within each subgraph has completely balanced out, but the two subgraphs have different temperature, because there is no edge between them, Fig. 2 (right). If one now plots the seven points in a 1D space according to their temperature, Fig. 2 (bottom right), all points of one subgraph cluster at one value and the points of the other subgraph cluster at another value. Thus, in this space separating the two subgraphs is trivial.

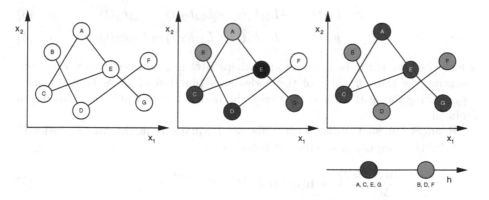

Fig. 2. Heat diffusion analogy of spectral clustering.

This is essentially how spectral clustering works. In real data the clusters, i.e. subgraphs, might not be completely disconnected, but with some tricks one can also deal with that.

The first step in spectral clustering is the Laplacian eigenmaps algorithm applied under the assumption that the graph consists of disconnected subgraphs and therefore results in clusters in the embedding. The second step in spectral clustering then discovers these clusters and partitions the nodes accordingly.

The graphs in Figs. 1 and 2 in the x_1-x_2-plane (not the embedding on the h-axis, of course) are drawn in a way that the position of the nodes actually has no meaning at all. This is to emphasize that the edges are the only thing that matters for the result of Laplacian eigenmaps and spectral clustering. In real world examples, however, spatial proximity often plays an important role and edges are preferably inserted between neighboring data points.

1.3   Heat Diffusion Equation for Connected Heat Reservoirs

How can we model heat diffusion mathematically, and how can we figure out the relevant temperature distributions analytically? Heat diffusion is a continuous process, so we need a differential equation (DE) for it. Since we consider heat diffusion between a discrete set of heat reservoirs rather than on a continuous medium, the DE is a system of ordinary DEs and not a partial DE. It is linear, e.g. if you have twice as much heat, diffusion will be twice as strong. And it is homogeneous, because if there is no heat, then there is no diffusion. Thus we consider the following system of ordinary linear DEs

$$\dot{\mathbf{h}}(t) = -\mathbf{L}\mathbf{h}(t) \tag{1}$$

$$\Longleftrightarrow \quad \dot{h}_1(t) = -L_{11}h_1(t) - L_{12}h_2(t) - L_{13}h_3(t) \tag{2}$$

$$\wedge \quad \dot{h}_2(t) = -L_{21}h_1(t) - L_{22}h_2(t) - L_{23}h_3(t) \tag{3}$$

$$\wedge \quad \dot{h}_3(t) = -L_{31}h_1(t) - L_{32}h_2(t) - L_{33}h_3(t) \tag{4}$$

spelled out for three heat reservoirs, where $\mathbf{h}(t)$ is a nonnegative vector representing the temperatures of the nodes as a function of time. $\mathbf{L}$ is a matrix representing the heat diffusion between the nodes, and it will be explained in a moment.

Readers not so familiar with differential equations might find it easier to consider the temporally discretized version of it,

$$\frac{\mathbf{h}(t + \Delta t) - \mathbf{h}(t)}{\Delta t} = \dot{\mathbf{h}}(t) \quad (\text{for } \Delta t \to 0) \tag{5}$$

$$\overset{(1)}{=} -\mathbf{L}\mathbf{h}(t) \tag{6}$$

$$\Longleftrightarrow \quad \mathbf{h}(t + \Delta t) = \mathbf{h}(t) - \Delta t\,\mathbf{L}\mathbf{h}(t) \tag{7}$$

$$= (\mathbf{I} - \Delta t\,\mathbf{L})\mathbf{h}(t) \quad (\text{with identity matrix } \mathbf{I}) \tag{8}$$

$$\Longleftrightarrow \quad h_1(t + \Delta t) = h_1(t) - \Delta t(L_{11}h_1(t) + L_{12}h_2(t) + L_{13}h_3(t)) \tag{9}$$

$$\wedge \quad h_2(t + \Delta t) = h_2(t) - \Delta t(L_{21}h_1(t) + L_{22}h_2(t) + L_{23}h_3(t)) \tag{10}$$

$$\wedge \quad h_3(t + \Delta t) = h_3(t) - \Delta t(L_{31}h_1(t) + L_{32}h_2(t) + L_{33}h_3(t)) \tag{11}$$

which is an approximation of the differential equation $\dot{\mathbf{h}}(t) = -\mathbf{L}\mathbf{h}(t)$, which is exact for $\Delta t \to 0$.

1.4   Laplacian Matrix

In either case, it is clear that $\mathbf{L}$ is responsible for any change of $\mathbf{h}$ and that the physics of the heat diffusion process imposes constraints on $\mathbf{L}$. If $\mathbf{L} = \mathbf{0}$ then $\mathbf{h}(t)$ is constant, which would correspond to three disconnected nodes (= heat reservoirs) that do not exchange any heat. A negative L_{ij} indicates that h_i increases proportional to h_j with factor $-L_{ij}$. A positive L_{ij} indicates that h_i decreases proportional to h_j with factor $-L_{ij}$.

We want that no heat gets lost or added to the system, thus $\sum_i L_{ij} = 0$ must be fulfilled, as one can easily verify by setting $\dot{h}_1(t) + \dot{h}_2(t) + \dot{h}_3(t) = 0$ or $h_1(t + \Delta t) + h_2(t + \Delta t) + h_3(t + \Delta t) = \text{const}$ for any values of $h_1(t), h_2(t)$, and $h_3(t)$. Since the heat one node gains must come from some other nodes, one can say that $-L_{ij}h_j$ (with negative L_{ij}) indicates the amount of heat node i gains from node j for $i \neq j$. The term $-L_{jj}h_j$ (with positive L_{jj}) indicates how much heat node j looses to the other nodes.

If we consider the situation that all three nodes are connected and one node, say Node 1, is hot and the other two nodes are absolutely freezing, i.e. $h_2 = h_3 = 0$ (Kelvin not Celsius) then initially only L_{11}, L_{21}, and L_{31} matter. It is intuitively clear that in this situation heat diffuses from Node 1 to Nodes 2 and 3, i.e. h_1 decreases and h_2 as well as h_3 increase proportionally to h_1. This implies $0 < L_{11}$, indicating that Node 1 looses heat, and $L_{21}, L_{31} < 0$, indicating that Nodes 2 and 3 gain heat from Node 1. If a connection would be absent, e.g. between Nodes 2 and 1, then no heat diffuses between these two nodes and the corresponding entry is zero, $L_{21} = 0$. If a node, let say Node 1, is not connected to any other node, then it cannot gain or loose heat at all, resulting in $L_{11} = 0$. Thus, by symmetry arguments we have $0 \leq L_{ii}$ and $L_{ij} \leq 0 \ \forall j \neq i$.

Finally, it is clear that if two different nodes i and j have same temperature, $h_i = h_j$, then the heat $-L_{ij}h_j$ diffusing from node j to node i equals the heat $-L_{ji}h_i$ diffusing from node i to node j, because otherwise one node would spontaneously become warmer and the other cooler, which would allow us to build a perpetual mobile. This implies $L_{ij} = L_{ji}$. Please notice here that if two connected nodes have same temperature, it does not mean that no heat diffuses from one to the other, it only means that the heat flows cancel out each other.

If we summarize the insights above we find that

$$L_{ij} = L_{ji} \qquad \text{(L is symmetric)} \qquad (12)$$

$$\sum_i L_{ij} \overset{(12)}{=} \sum_j L_{ij} = 0 \qquad \text{(rows and columns add up to zero)} \quad (13)$$

$$L_{ii} \geq 0 \qquad \text{(diagonal elements are non-negative)} \quad (14)$$

$$L_{ij} \leq 0 \quad \forall j \neq i \quad \text{(off-diagonal elements are non-positive)} \tag{15}$$

An example of a matrix with all these properties is

$$\mathbf{L} = \begin{pmatrix} 0.2 & -0.2 & 0 \\ -0.2 & 1.0 & -0.8 \\ 0 & -0.8 & 0.8 \end{pmatrix} \tag{16}$$

The corresponding graph is shown in Fig. 3.

1.5 Solution of the Heat Diffusion Equation

For a given square matrix $\mathbf{L}$, the solutions to the so-called eigenvalue equation

$$\mathbf{L}\mathbf{u}_\alpha = \gamma_\alpha \mathbf{u}_\alpha \tag{17}$$

are called eigenvectors $\mathbf{u}_\alpha$ and eigenvalues γ_α. Normally, when multiplying a matrix with a vector the matrix changes length and direction of a vector. Eigenvectors are special in that they are changed only in length but not in direction, thus the effect of the matrix can be expressed by simply multiplying with a scalar, which is what Eq. (17) represents. Equations that are originally written with matrices can thus simplify significantly for eigenvectors, which makes it often very useful to represent normal vectors as linear combinations of eigenvectors.

Assume the eigenvectors $\mathbf{u}_\alpha$ and eigenvalues γ_α of the Laplacian matrix are known and ordered such that $\gamma_1 \leq \gamma_2 \leq \dots \leq \gamma_I$. It turns out that all eigenvalues are non-negative (Property $\langle 4 \rangle$ in Sect. 2.8) and from (13) follows directly that one can chose $\mathbf{u}_1 = (1, 1, \dots, 1)^T$ (usually normalized to norm one by convention) with $\gamma_1 = 0$ as the first eigenvector and -value (Property $\langle 7 \rangle$ in Sect. 2.8).

Because the Laplacian matrix is symmetric and real, the set of eigenvectors is complete and forms a basis for the vector space. Any initial temperature vector $\mathbf{h}(t = 0)$ can thus be written as a linear combination of the eigenvectors

$$\mathbf{h}(t = 0) = \sum_\alpha \omega_\alpha \mathbf{u}_\alpha \tag{18}$$

with some appropriate prefactors ω_α.

From the theory of systems of homogeneous linear differential equations we know that the general solution of (1) for this $\mathbf{h}(t = 0)$ is then

$$\mathbf{h}(t) = \sum_\alpha \omega_\alpha \exp(-\gamma_\alpha t) \mathbf{u}_\alpha \tag{19}$$

which, for non-negative eigenvalues γ_α, is a superposition of eigenvectors with exponentially decaying weights.

For the discretized version of the differential equation we first observe that

$$\underbrace{(\mathbf{I} - \Delta t \mathbf{L})}_{=:\mathbf{P}} \mathbf{u}_\alpha = (\mathbf{I} \mathbf{u}_\alpha - \Delta t \mathbf{L} \mathbf{u}_\alpha) \tag{20}$$

$$\overset{(17)}{=} (\mathbf{u}_\alpha - \Delta t \gamma_\alpha \mathbf{u}_\alpha) \tag{21}$$

$$= \underbrace{(1 - \Delta t \gamma_\alpha)}_{=:\xi_\alpha} \mathbf{u}_\alpha \tag{22}$$

$$\Longleftrightarrow \qquad \mathbf{P} \mathbf{u}_\alpha = \xi_\alpha \mathbf{u}_\alpha \tag{23}$$

Thus the $\mathbf{u}_\alpha$ are also eigenvectors of $\mathbf{P}$ but with eigenvalues $\xi_\alpha = (1 - \Delta t \gamma_\alpha)$ with $1 = \xi_1 \geq \xi_2 \geq ... \geq \xi_I > 0$ for small enough Δt. With this we find

$$\mathbf{h}(t = N\Delta t) \stackrel{(8,20)}{=} \mathbf{P}^N \mathbf{h}(0) \tag{24}$$

$$\stackrel{(18)}{=} \mathbf{P}^N \sum_\alpha \omega_\alpha \mathbf{u}_\alpha \tag{25}$$

$$= \sum_\alpha \omega_\alpha \mathbf{P}^N \mathbf{u}_\alpha \tag{26}$$

$$\stackrel{(23)}{=} \sum_\alpha \omega_\alpha \xi_\alpha^N \mathbf{u}_\alpha \tag{27}$$

which, for eigenvalues ξ_α between zero and one, again is a superposition of eigenvectors with exponentially decaying weights.

In either case, if one waits long enough, only the first eigenvectors with eigenvalue $\gamma_\alpha = 0$ respectively $\xi_\alpha = 1$ will still contribute to $\mathbf{h}(t)$, and one can show that if the graph is connected, only the contribution of $\mathbf{u}_1$ survives indefinitely long, because $\exp(-\gamma_1 t) = \exp(-0t) = 1$ and $\xi_1^N = 1^N = 1$ for any t. The last eigenvector fading away is $\mathbf{u}_2$, and that is exactly the vector we are interested in for the Laplacian eigenmaps algorithm, see Fig. 1 (right).

If the graph is disconnected then it is intuitively clear that each subgraph balances out its heat over time, but there is no heat exchange between subgraphs. The corresponding Laplacian matrix becomes a block matrix with as many blocks on the diagonal as there are subgraphs. In the example above in Fig. 2, there are two subgraphs, and because of the block structure of the Laplacian matrix and the fact that rows add up to zero, one can verify that the second eigenvector $\mathbf{u}_2 = (1/4, -1/3, 1/4, -1/3, 1/4, -1/3, 1/4)^T$ (usually normalized to norm one by convention) is constant within each subgraph and has eigenvalue $\gamma_2 = 0$. This again reflects the temperature distribution that remains if one waits for a long time, and that is exactly the vector we are interested in the spectral clustering algorithm, see Fig. 2 (right).

In summary, the second eigenvector of the Laplacian matrix provides a nice 1D arrangement of the nodes of a similarity graph. In practice one often also uses the third and possibly the forth eigenvector to get visualizations in 2D or 3D, but that is not so easy to understand with this intuitive explanation.

2   Formalism

After the intuitive explanation we now consider Laplacian eigenmaps and spectral clustering more directly and more formally. For both algorithms data must first be represented as a graph. Nodes represent data samples and edges represent similarities between data samples. The samples could be anything, e.g. words, persons, or melodies, they need not be vectors in a vector space. We just need a non-negative function that measures similarity between two data samples. And this function does not even need to be consistent with a metric, i.e. does not

need to fulfill the triangle inequality stating that the distance between A and C cannot be greater than the sum of distances from A to B and B to C, where A, B, and C are arbitrary points in space. We first introduce some notions from graph theory and then consider the optimization problem.

2.1   Simple Graphs

A *graph* $G = (\mathbb{V}, \mathbb{E})$ is a set of *nodes* (or vertices or points) $\mathbb{V} = \{v_1, ..., v_I\}$ and a set of *edges* $\mathbb{E} = \{e_1, ..., e_L\}$. An edge e_l connects two nodes v_i and v_j and is therefore defined by a pair of nodes. Edges may be *directed*, going from node v_i to node v_j, indicated by $e_l = (v_i, v_j)$. Edges may also be *undirected*, in which case the order of the nodes does not matter and we can write $e_l := \{v_i, v_j\}$, where the curly brackets imply that the order does not matter. *Simple graphs* are undirected graphs without loops, which are edges that connect a node with itself, and no parallel edges, which are edges that connect the same pair of nodes. Here we consider mainly simple graphs.

Further reading: [8].

2.2   Matrix Representation

Graphs can be conveniently represented by real matrices. The *adjacency matrix* $\mathbf{A} = (A_{ij})$ of an undirected graph is $I \times I$ and defined as

$$A_{ij} := \begin{cases} 1 & \text{if } \{v_i, v_j\} \in \mathbb{E} \\ 0 & \text{otherwise} \end{cases} \tag{28}$$

i.e. it has a one in entry A_{ij} if and only if nodes v_i and v_j are connected with each other. Matrix $\mathbf{A}$ is naturally symmetric, since the edges are not directed.

The *degree matrix* $\mathbf{D} = (D_{ij})$ of an undirected graph is a diagonal matrix, where the diagonal entries D_{ii} indicate the number of edges connected to node v_i.

In context of the Laplacian matrix, we generalize these definitions to weighted graphs, where the edges are labeled with a real (positive) number indicating their weight W_{ij}. If one simply replaces the 1 values in (28) by these weights, then $\mathbf{A}$ becomes the (edge) weight matrix $\mathbf{W}$, and the weighted degree matrix $\mathbf{D} = (D_{ij})$ gets the sum over all weights of the edges converging on a node in their diagonal entries.

$$D_{ii} := \sum_j W_{ij} = \sum_j W_{ji} \tag{29}$$

Figure 3 shows a simple weighted graph. The weighted adjacency matrix, or weight matrix, of the undirected graph is

$$\mathbf{W} = \begin{pmatrix} & v_1 & v_2 & v_3 \\ v_1 & 0 & 0.2 & 0 \\ v_2 & 0.2 & 0 & 0.8 \\ v_3 & 0 & 0.8 & 0 \end{pmatrix} \tag{30}$$

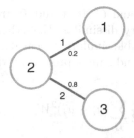

Fig. 3. Example of a simple, weighted, undirected graph. Edges are numbered with blue (or gray) integers, their real valued weights are shown in black. Weights do not need to add up to one, like here for Node 2. (Color figure online)

The weighted degree matrix of the undirected graph is

$$\mathbf{D} = \begin{pmatrix} & v_1 & v_2 & v_3 \\ v_1 & 0.2 & 0 & 0 \\ v_2 & 0 & 1.0 & 0 \\ v_3 & 0 & 0 & 0.8 \end{pmatrix} \tag{31}$$

The *Laplacian matrix* $\mathbf{L}$ is defined as the difference between weighted degree matrix $\mathbf{D}$ and weight matrix $\mathbf{W}$

$$\mathbf{L} = \mathbf{D} - \mathbf{W}. \tag{32}$$

It is easy to verify that it has all the properties (12–15) derived in Sect. 1.4 from the heat diffusion analogy.

The Laplacian matrix for the example above is

$$\mathbf{L} = \begin{pmatrix} 0.2 & -0.2 & 0 \\ -0.2 & 1.0 & -0.8 \\ 0 & -0.8 & 0.8 \end{pmatrix} \tag{33}$$

2.3 Optimization Problem

The objective of Laplacian eigenmaps as well as spectral clustering is to assign[2] similar values to similar nodes, i.e. strongly connected nodes, and dissimilar values to nodes that are not similar. This is a non-trivial operation, since similarity is a property of a pair of nodes, or an edge, while value is a property of a single node. It is not guaranteed that there is a good solution at all. Consider, for

---

[2] A remark on terminology: We use *assign/assignment* for giving data samples an associated value. These values implicitly define a *mapping* from (possibly high-dimensional or non-vectorial) data samples to points in a low-dimensional space, the *mapped space*. In LPP the mapping is defined more explicitly by a linear function. The collection of points in mapped space form an *embedding*. Thus, all these terms refer to the same process.

instance, three nodes A, B, and C. If A and B are very similar as well as B and C, but A and C are very dissimilar, then there are no values that could reflect that. However, reasonable similarity measures usually do not lead to such conflicts, definitely not those inducing a proper metric. In any case, the objective is to

$$\text{minimize} \quad \frac{1}{2}\sum_{ij}(u_i - u_j)^2 W_{ij} \tag{34}$$

$$\text{subject to} \quad \mathbf{1}^T\mathbf{u} = 0 \quad \text{(zero mean)} \tag{35}$$

$$\text{and} \quad \mathbf{u}^T\mathbf{u} = 1 \quad \text{(unit variance)} \tag{36}$$

$$\text{or subject to} \quad \mathbf{1}^T\mathbf{D}\mathbf{u} = 0 \quad \text{(weighted zero mean)} \tag{37}$$

$$\text{and} \quad \mathbf{u}^T\mathbf{D}\mathbf{u} = 1 \quad \text{(weighted unit variance)} \tag{38}$$

with $\mathbf{u} = (u_1, u_2, ..., u_I)^T$ and $\mathbf{1} = (1, 1, 1, ..., 1)^T$ indicating the one-vector. Objective (34) favors solutions where strongly connected nodes with a large edge weight W_{ij} have similar values u_i and u_j. Constraints (35) and (36) in conjunction avoid the trivial constant solution, which implicitly guarantees that nodes that are not similar get assigned dissimilar values. Constraints (37) and (38) have the same function but imply some normalization, see Sect. 2.5.

If we need more than one solution in order to map the nodes into a higher dimensional space, we add a subscript index to $\mathbf{u}$ and solve the same optimization problem multiple times subject to the additional constraint

$$\mathbf{u}_\beta^T\mathbf{u}_\alpha = 0 \quad \forall \beta < \alpha \quad \text{(decorrelation to previous solutions)} \tag{39}$$

$$\text{or} \quad \mathbf{u}_\beta^T\mathbf{D}\mathbf{u}_\alpha = 0 \quad \forall \beta < \alpha \quad \text{(decorrelation to previous solutions)} \tag{40}$$

for the second and later solutions $\mathbf{u}_\alpha$ to make them different (orthogonal) to the previous solutions $\mathbf{u}_\beta$.

2.4   Associated Eigenvalue Problem

It is known that the normalized eigenvectors $\mathbf{u}_\alpha$ of the ordinary eigenvalue equation

$$\mathbf{L}\mathbf{u}_\alpha = \gamma_\alpha \mathbf{u}_\alpha \tag{41}$$

ordered by increasing eigenvalues γ_α solve the optimization problem

$$\text{minimize} \quad \mathbf{u}_\alpha^T\mathbf{L}\mathbf{u}_\alpha = \frac{1}{2}\sum_{ij}(u_{\alpha,i} - u_{\alpha,j})^2 W_{ij} \tag{42}$$

$$\text{subject to} \quad \mathbf{u}_\alpha^T\mathbf{u}_\alpha = 1 \quad \text{(unit norm)} \tag{43}$$

$$\text{and} \quad \mathbf{u}_\beta^T\mathbf{u}_\alpha = 0 \quad \forall \beta < \alpha \quad \text{(order and orthogonality)} \tag{44}$$

where constraint (44) induces an order such that $\mathbf{u}_1$ is the optimal solution without any orthogonality constraint (only the unit norm constraint), $\mathbf{u}_2$ is the

optimal solution with the additional constraint of being orthogonal to $\mathbf{u}_1$, $\mathbf{u}_3$ is the optimal solution with the additional constraint of being orthogonal to $\mathbf{u}_1$ and $\mathbf{u}_2$, etc. Constraints (43, 44) can be combined to $\mathbf{u}_\beta^T \mathbf{u}_\alpha = \delta_{\beta\alpha} \ \forall \beta \leq \alpha$. Identity (42) is left to the reader as an exercise. If one orders the eigenvalues by ascending rather than descending value, the corresponding eigenvectors solve the maximization rather than minimization problem.

The zero mean constraint (35) is implicit here. Since the first solution $\mathbf{u}_1$ is a scaled version of $\mathbf{1}$, Constraint (44) with $\beta = 1$ is equivalent to (35). The solutions of interest thus start with index 2 rather than 1.

Since

$$\mathbf{u}_\alpha^T \mathbf{L} \mathbf{u}_\alpha \overset{(41)}{=} \mathbf{u}_\alpha^T \gamma_\alpha \mathbf{u}_\alpha = \gamma_\alpha \mathbf{u}_\alpha^T \mathbf{u}_\alpha \overset{(43)}{=} \gamma_\alpha \tag{45}$$

the eigenvalues are the optimal values of the objective function.

In the algorithms below the constraint is usually $\mathbf{w}^T \mathbf{D} \mathbf{w} = 1$ rather than $\mathbf{u}^T \mathbf{u} = 1$ (we switch here from $\mathbf{u}$ to $\mathbf{w}$ to indicate solutions with this weighted normalization). Thus we note that the appropriately normalized eigenvectors $\mathbf{w}_\alpha$ of the generalized eigenvalue equation

$$\mathbf{L} \mathbf{w}_\alpha = \lambda_\alpha \mathbf{D} \mathbf{w}_\alpha \tag{46}$$

ordered by increasing eigenvalues λ_α solve the optimization problem

$$\text{minimize} \qquad \mathbf{w}_\alpha^T \mathbf{L} \mathbf{w}_\alpha = \frac{1}{2} \sum_{ij} (w_{\alpha,i} - w_{\alpha,j})^2 W_{ij} \tag{47}$$

$$\text{subject to} \qquad \mathbf{w}_\alpha^T \mathbf{D} \mathbf{w}_\alpha = 1 \qquad \text{(weighted unit norm)} \tag{48}$$

$$\text{and} \qquad \mathbf{w}_\beta^T \mathbf{D} \mathbf{w}_\alpha = 0 \quad \forall \beta < \alpha \qquad \text{(order and weighted orthogonality)} \tag{49}$$

The derivation (45) does not hold here, since the eigenvectors must have weighted unit norm, not standard unit norm. But still we find analogously

$$\mathbf{w}_\alpha^T \mathbf{L} \mathbf{w}_\alpha \overset{(46)}{=} \mathbf{w}_\alpha^T \lambda_\alpha \mathbf{D} \mathbf{w}_\alpha = \lambda_\alpha \mathbf{w}_\alpha^T \mathbf{D} \mathbf{w}_\alpha \overset{(48)}{=} \lambda_\alpha \tag{50}$$

Thus, in both cases the eigenvalues are the value of the objective function for the different eigenvectors. Eigenvectors with small eigenvalue are smooth in the sense that connected nodes tend to have similar values while eigenvectors with large eigenvalue are more rugged, i.e. connected nodes tend to have different values, see Eqs. (42, 47) and Fig. 4.

Further reading: [10].

2.5   The Role of the Weighted Normalization Constraint

What is the difference between the constraints $\mathbf{u}_\alpha^T \mathbf{u}_\alpha = 1$ (43) and $\mathbf{w}_\alpha^T \mathbf{D} \mathbf{w}_\alpha = 1$ (48)? Since $\mathbf{D}$ is a diagonal matrix, this simply means that in the constraint the components of the generalized eigenvectors get weighted by $\sqrt{D_{ii}}$ (29) (the square root comes from the fact that in $\mathbf{w}_\alpha^T \mathbf{D} \mathbf{w}_\alpha$ the D_{ii} has to be equally distributed over the two $\mathbf{w}_\alpha$). For the term $w_i D_{ii} w_i$ to have the same effect size

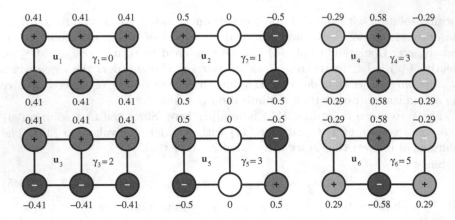

Fig. 4. Example of eigenvectors and -values for a graph with six nodes for the optimization problem with standard norm (34–36) corresponding to the ordinary eigenvalue problem (41). Eigenvectors $\mathbf{u}_\alpha$ with small eigenvalue γ_α tend to have similar values for connected nodes and are correspondingly smooth; those with large eigenvalues tend to have different values for connected nodes and are correspondingly rugged. Notice that $\mathbf{u}_4$ and $\mathbf{u}_5$ have the same eigenvalue 3, so that any linear combination of them is also an eigenvector with eigenvalue 3. Thus, these eigenvectors are not unique.

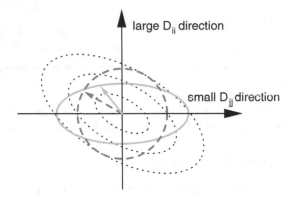

Fig. 5. Visualization of the role of the constraint on the optimization problem. The dotted ellipses illustrate the quadratic form being minimized (42, 47), which is the same for both problems. The blue dashed circle and green solid ellipse illustrate the constraints (43) and (48), respectively. The corresponding arrow indicates the optimal solution, which is the point on the circle or ellipse that comes closest to the inner dashed ellipses. (Color figure online)

in the constraint, a component w_i with large D_{ii} must be smaller than one with a small D_{ii}. This is illustrated in Fig. 5 by the green solid ellipse vs the blue dashed circle. The latter is the set of points with $\mathbf{u}_\alpha^T\mathbf{u}_\alpha = 1$, the former the set with $\mathbf{w}_\alpha^T\mathbf{D}\mathbf{w}_\alpha = 1$ with large D_{ii} and small D_{jj}.

In the figure it is assumed that the determinant of $\mathbf{D}$ is one. That does not need to be the case. It could be any other positive value, depending on how strong the weights of the edges are. However, a consistent scaling of the weights does not change the solution, so we can assume w.l.o.g. that they are scaled such that $|\mathbf{D}| = 1$.

While the constraint differs, the objective function (42, 47) is the same in both cases. It takes the form of an unisotropic paraboloid, like a squeezed champagne glass, indicated in Fig. 5 by dotted ellipses. Minimizing it under the constraint means finding the point on the blue circle or green ellipse that comes closest to the inner ellipses. To the extent the D_{ii} differ, the components with larger D_{ii} are favored over components with smaller D_{ii}, because they allow the vector $\mathbf{w}_\alpha$ to move closer to the origin, where the true minimum of the objective function with value 0 lies.

However, this does not mean that all components of $\mathbf{w}_\alpha$ with large D_{ii} become larger relative to those with small D_{ii}. That depends also on the objective function. But the general tendency is that the change from constraint $\mathbf{u}_\alpha^T \mathbf{u}_\alpha = 1$ to constraint $\mathbf{w}_\alpha^T \mathbf{D} \mathbf{w}_\alpha = 1$ makes the values of highly connected nodes (with large D_{ii}) larger relative to less connected nodes (with small D_{ii}).

Why might that be useful? Imagine a square lattice of 7×7 nodes, connected with their four nearest neighbors with equal edge weights one. This looks like a pretty good connectivity to represent the 2D layout of the grid. Now, imagine in the right half of the grid, each node is connected to its eight nearest neighbors instead of four. Both, the four- as well as the eight-neighbor connectivity, are perfectly fine representations of the 2D layout. But because the nodes on the right side have more edges, heat would diffuse faster and temperature would equalize more quickly, leading to more similar values, the nodes would move closer together in the embedding. If one uses constraint $\mathbf{w}_\alpha^T \mathbf{D} \mathbf{w}_\alpha = 1$ this advantage of the more densely connected half would be somewhat compensated by scaling up the values, which also leads to larger differences. This leads to a value distribution that better reflects the 2D layout and is less influenced by the different density of connections between left and right half.

It is probably also possible to construct examples where the constraint $\mathbf{u}_\alpha^T \mathbf{u}_\alpha = 1$ gives more desirable results. But at least it should be clear now what the effect of the constraint $\mathbf{w}_\alpha^T \mathbf{D} \mathbf{w}_\alpha = 1$ is, it somewhat counteracts the effect of systematically strong (or weak) connections in a region of the graph. This does not tell much about the effects on a more microscopic level. But it is clear that it makes no sense to change the value of a single highly connected node and make it too different from the values of its neighbors, because that really contributes to a bad value in the objective function.

2.6 Symmetric Normalized Laplacian Matrix

For the algorithms below, we consider the eigenvalues and -vectors of the generalized eigenvalue equation $\mathbf{L} \mathbf{w}_\alpha = \lambda_\alpha \mathbf{D} \mathbf{w}_\alpha$. Since most of us are more familiar with the ordinary eigenvalue equation, it is interesting to note that one can convert the generalized eigenvalue equation into an ordinary one and back again.

This allows us to transfer what we know about ordinary eigenvalue equations to the generalized ones.

First assume $D_{ii} \neq 0 \; \forall i$ ($0 \leq D_{ii}$ is true in any case) and define

$$\mathbf{d} := (D_{11}, ..., D_{II})^T \tag{51}$$

$$\overline{\mathbf{d}} := (\sqrt{D_{11}}, ..., \sqrt{D_{II}})^T \tag{52}$$

$$\underline{\mathbf{d}} := (1/\sqrt{D_{11}}, ..., 1/\sqrt{D_{II}})^T \tag{53}$$

$$\mathbf{D} := \text{diag}(\mathbf{d}) = \mathbf{D}^T \tag{54}$$

$$\overline{\mathbf{D}} := \text{diag}(\overline{\mathbf{d}}) = \overline{\mathbf{D}}^T \tag{55}$$

$$\underline{\mathbf{D}} := \text{diag}(\underline{\mathbf{d}}) = \underline{\mathbf{D}}^T \tag{56}$$

so that, for instance, $\overline{\mathbf{D}}\,\underline{\mathbf{D}} = \underline{\mathbf{D}}\,\overline{\mathbf{D}} = \mathbf{I}$ and $\overline{\mathbf{D}}\,\overline{\mathbf{D}} = \mathbf{D}$.

Now we convert the generalized eigenvalue equation into an ordinary one.

$$\mathbf{L}\mathbf{w}_\alpha \overset{!}{=} \lambda_\alpha \mathbf{D}\mathbf{w}_\alpha \quad | \underline{\mathbf{D}}\cdot \tag{57}$$

$$\Longleftrightarrow \quad \underline{\mathbf{D}}\mathbf{L}\,\underbrace{\underline{\mathbf{D}}\,\overline{\mathbf{D}}}_{=\mathbf{I}}\mathbf{w}_\alpha = \underline{\mathbf{D}}\lambda_\alpha\underbrace{\overline{\mathbf{D}}\,\overline{\mathbf{D}}}_{=\mathbf{D}}\mathbf{w}_\alpha \quad (\text{since } \underline{\mathbf{D}} \text{ is invertible}) \tag{58}$$

$$\Longleftrightarrow \quad \underbrace{\underline{\mathbf{D}}\mathbf{L}\underline{\mathbf{D}}}_{=:\hat{\mathbf{L}}}\,\underbrace{\overline{\mathbf{D}}\mathbf{w}_\alpha}_{=:\hat{\mathbf{w}}_\alpha} = \lambda_\alpha \underbrace{\overline{\mathbf{D}}\,\overline{\mathbf{D}}}_{=\mathbf{I}}\,\underbrace{\overline{\mathbf{D}}\mathbf{w}_\alpha}_{=:\hat{\mathbf{w}}_\alpha} \tag{59}$$

$$\Longleftrightarrow \quad \hat{\mathbf{L}}\hat{\mathbf{w}}_\alpha = \lambda_\alpha\hat{\mathbf{w}}_\alpha \tag{60}$$

with

$$\hat{\mathbf{w}}_\alpha = \overline{\mathbf{D}}\mathbf{w}_\alpha \tag{61}$$

$$\Longleftrightarrow \quad \mathbf{w}_\alpha = \underline{\mathbf{D}}\hat{\mathbf{w}}_\alpha \tag{62}$$

and the *symmetric normalized Laplacian matrix*

$$\hat{\mathbf{L}} := \underline{\mathbf{D}}\mathbf{L}\underline{\mathbf{D}} \tag{63}$$

Thus, if and only if $\mathbf{w}_\alpha$ is an eigenvector of the generalized eigenvalue equation with eigenvalue λ_α, then $\hat{\mathbf{w}}_\alpha$ is an eigenvector of the ordinary eigenvalue equation with same eigenvalue λ_α. It is sometimes helpful to switch back and forth between these two views.

For the example above we find

$$\hat{\mathbf{L}} = \begin{pmatrix} & \div\sqrt{0.2} & \div\sqrt{1.0} & \div\sqrt{0.8} \\ & \downarrow & \downarrow & \downarrow \\ \div\sqrt{0.2} \rightarrow & 0.2 & -0.2 & 0 \\ \div\sqrt{1.0} \rightarrow & -0.2 & 1.0 & -0.8 \\ \div\sqrt{0.8} \rightarrow & 0 & -0.8 & 0.8 \end{pmatrix} = \begin{pmatrix} 1.0 & -\sqrt{0.2} & 0 \\ -\sqrt{0.2} & 1.0 & -\sqrt{0.8} \\ 0 & -\sqrt{0.8} & 1.0 \end{pmatrix} \tag{64}$$

where $\div\sqrt{\cdot}$ indicates multiplication with $\underline{\mathbf{D}}$ from the left along the rows and from the right along the columns. It is easy to see that $\hat{L}_{ii} = 1$ by construction,

since $\underline{\mathbf{D}}\mathbf{L}\underline{\mathbf{D}} = \underline{\mathbf{D}}(\mathbf{D} - \mathbf{W})\underline{\mathbf{D}} = (\mathbf{I} - \underline{\mathbf{D}}\mathbf{W}\underline{\mathbf{D}})$ and $\underline{\mathbf{D}}\mathbf{W}\underline{\mathbf{D}}$ has only zeroes on the diagonal. But the rows and columns do not add up to zero anymore.

The objective function related to the eigenvalue equation of the symmetric normalized Laplacian matrix is

$$\hat{\mathbf{w}}_\alpha^T \hat{\mathbf{L}}\hat{\mathbf{w}}_\alpha \stackrel{(63)}{=} \hat{\mathbf{w}}_\alpha^T \underline{\mathbf{D}}\mathbf{L}\underline{\mathbf{D}}\hat{\mathbf{w}}_\alpha \tag{65}$$

$$= (\underline{\mathbf{D}}\hat{\mathbf{w}}_\alpha)^T \mathbf{L}\underline{\mathbf{D}}\hat{\mathbf{w}}_\alpha \quad (\text{since } \underline{\mathbf{D}} \text{ is diagonal, thus } \underline{\mathbf{D}} = \underline{\mathbf{D}}^T) \tag{66}$$

$$\stackrel{(42)}{=} \frac{1}{2}\sum_{ij}((\underline{\mathbf{D}}\hat{\mathbf{w}}_\alpha)_i - (\underline{\mathbf{D}}\hat{\mathbf{w}}_\alpha)_j)^2 W_{ij} \tag{67}$$

$$\stackrel{(56,53)}{=} \frac{1}{2}\sum_{ij}\left(\frac{\hat{w}_{\alpha,i}}{\sqrt{D_{ii}}} - \frac{\hat{w}_{\alpha,j}}{\sqrt{D_{jj}}}\right)^2 W_{ij} \quad (\text{since } \underline{\mathbf{D}} \text{ is diagonal}) \tag{68}$$

2.7 Random Walk Normalized Laplacian Matrix

Another possibility to convert the generalized eigenvalue equation into an ordinary one is simply to multiply (46) from the left with the inverse of the weighted degree matrix.

$$\mathbf{L}\mathbf{w}_\alpha \stackrel{(46)}{=} \lambda_\alpha \mathbf{D}\mathbf{w}_\alpha \quad | \ \mathbf{D}^{-1}. \tag{69}$$

$$\Longleftrightarrow \underbrace{\mathbf{D}^{-1}\mathbf{L}}_{=:\hat{\mathbf{L}}^{\mathrm{rw}}}\mathbf{w}_\alpha = \lambda_\alpha\mathbf{w}_\alpha \quad (\text{since } \mathbf{D} \text{ is invertible}) \tag{70}$$

$$\Longleftrightarrow \hat{\mathbf{L}}^{\mathrm{rw}}\mathbf{w}_\alpha = \lambda_\alpha\mathbf{w}_\alpha \tag{71}$$

$\hat{\mathbf{L}}^{\mathrm{rw}} := \mathbf{D}^{-1}\mathbf{L}$ is the *random walk normalized Laplacian matrix* and has the same eigenvalues and eigenvectors as the generalized eigenvalue equation of the Laplacian matrix. Its main disadvantage is that it is non-symmetric.

For the example above we find

$$\hat{\mathbf{L}}^{\mathrm{rw}} = \begin{pmatrix} \div\,0.2 \to & 0.2 & -0.2 & 0 \\ \div\,1.0 \to & -0.2 & 1.0 & -0.8 \\ \div\,0.8 \to & 0 & -0.8 & 0.8 \end{pmatrix} = \begin{pmatrix} 1.0 & -1.0 & 0 \\ -0.2 & 1.0 & -0.8 \\ 0 & -1.0 & 1.0 \end{pmatrix} \tag{72}$$

where $\div\cdot$ indicates multiplication with $\mathbf{D}^{-1}$ from the left along the rows. Notice that $\hat{L}_{ii}^{\mathrm{rw}} = 1$ and that the rows, but not the columns, add up to zero. $\mathbf{P} := \mathbf{I} - \hat{\mathbf{L}}^{\mathrm{rw}}$ is a *right stochastic matrix* [12], which can be interpreted as a transition matrix for a random walk between the nodes of the graph. Therefore the name. We are not sure how useful this intuition is, since the right stochastic matrix has to be multiplied from the right, in order to simulate a random walk, but in the eigenvalue equation $\hat{\mathbf{L}}^{\mathrm{rw}}$ is multiplied from the left.

In what follows we focus on $\hat{\mathbf{L}}$ rather than $\hat{\mathbf{L}}^{\mathrm{rw}}$, because the non-symmetry makes the latter more difficult to deal with.

2.8 Summary of Mathematical Properties

The Laplacian matrix appears in a multitude of different algorithms, three of which will be discussed in this lecture: *Laplacian eigenmaps (LEM)*, *locality preserving projections (LPP)*, and *spectral clustering*. When using the Laplacian matrix in an algorithm, we are usually interested in its eigenvectors and eigenvalues. The set of eigenvalues of a matrix is referred to as its *spectrum*.

The Laplacian matrix, its eigenvectors, and its spectrum have the following properties (see also Table 1):

1. $\mathbf{L}$ and $\hat{\mathbf{L}}$ are both symmetric (and real). The symmetry of $\mathbf{L}$ follows directly from Eq. (32) since $\mathbf{D}$ is diagonal and $\mathbf{W}$ is symmetric. The symmetry of $\hat{\mathbf{L}}$ follows from Eq. (63) and the symmetry of $\mathbf{L}$. See (33) and (64) for the example above.

2. $\mathbf{L}$ and $\hat{\mathbf{L}}$ each have a complete set of orthogonal eigenvectors $\mathbf{u}_\alpha$ and $\hat{\mathbf{w}}_\alpha$, respectively, with real eigenvalues. This is true for any real symmetric matrix, see Property $\langle 1 \rangle$.

3. $\mathbf{L}$ and $\hat{\mathbf{L}}$ are both positive semi-definite (meaning that $\mathbf{x}^T \mathbf{L} \mathbf{x} \geq 0$ for any vector $\mathbf{x}$). For $\mathbf{L}$ this follows directly from (42) and the fact that all weights are positive; for $\hat{\mathbf{L}}$ this follows from Eq. (63) and the fact that it holds for $\mathbf{L}$.

4. $\mathbf{L}$ and $\hat{\mathbf{L}}$ have only non-negative eigenvalues. This follows from Property $\langle 3 \rangle$. Note, however, that the eigenvalues of $\mathbf{L}$ and $\hat{\mathbf{L}}$ may be different. We indicate the eigenvalues of $\mathbf{L}$ by γ_α and those of $\hat{\mathbf{L}}$ by λ_α.

5. $\hat{\mathbf{L}}\hat{\mathbf{w}}_\alpha = \lambda_\alpha \hat{\mathbf{w}}_\alpha$ and $\mathbf{L}\mathbf{w}_\alpha = \lambda_\alpha \mathbf{D}\mathbf{w}_\alpha$ have the same set of eigenvalues λ_α and their eigenvectors are related by $\mathbf{w}_\alpha = \underline{\mathbf{D}}\hat{\mathbf{w}}_\alpha \Leftrightarrow \hat{\mathbf{w}}_\alpha = \overline{\mathbf{D}}\mathbf{w}_\alpha$, see Sect. 2.6.

6. The generalized eigenvalue equation $\mathbf{L}\mathbf{w}_\alpha = \lambda_\alpha \mathbf{D}\mathbf{w}_\alpha$ has only non-negative eigenvalues λ_α and a full set of eigenvectors $\mathbf{w}_\alpha$ that are orthogonal with respect to the inner product $\mathbf{w}_\beta \mathbf{D}\mathbf{w}_\alpha$ for $\beta \neq \alpha$. This follows from Properties $\langle 2,4 \rangle$ with Property $\langle 5 \rangle$, since $\forall \beta \neq \alpha : 0 \overset{(2)}{=} \hat{\mathbf{w}}_\beta^T \hat{\mathbf{w}}_\alpha \overset{(61)}{=} \mathbf{w}_\beta^T \overline{\mathbf{D}\mathbf{D}}\mathbf{w}_\alpha = \mathbf{w}_\beta \mathbf{D}\mathbf{w}_\alpha$.

7. $\mathbf{1} := (1,1,...,1)^T$ (the one-vector) is a solution of the ordinary eigenvalue equation $\mathbf{L}\mathbf{u}_\alpha = \gamma_\alpha \mathbf{u}_\alpha$ as well as the generalized eigenvalue equation $\mathbf{L}\mathbf{w}_\alpha = \lambda_\alpha \mathbf{D}\mathbf{w}_\alpha$ with eigenvalue 0. This follows directly from the definition of $\mathbf{L}$ (32), since its rows sum up to zero, so that $\mathbf{L}\mathbf{1} = \mathbf{0} = 0 \cdot \mathbf{1}$, and because the two eigenvalue equations are identical for $\gamma_\alpha = \lambda_\alpha = 0$. We chose the appropriately normalized one-vector to be the first eigenvectors $\mathbf{u}_1 = 1/\sqrt{\mathbf{1}^T \mathbf{1}}$ and $\mathbf{w}_1 = 1/\sqrt{\mathbf{1}^T \mathbf{D}\mathbf{1}}$ with $\gamma_1 = \lambda_1 = 0$.

8. $\overline{\mathbf{d}}$, see (52), is a solution of the ordinary eigenvalue equation $\hat{\mathbf{L}}\hat{\mathbf{w}}_\alpha = \lambda_\alpha \hat{\mathbf{w}}_\alpha$ with eigenvalue 0. This follows from Property $\langle 7 \rangle$ and Eq. (62) since $\underline{\mathbf{D}}\overline{\mathbf{d}} = \mathbf{1} \overset{\langle 7 \rangle}{=} \mathbf{w}_1$. We chose this 'square-root degree-vector' normalized to norm one to be the first eigenvector $\hat{\mathbf{w}}_1 = \overline{\mathbf{d}}/\sqrt{\overline{\mathbf{d}}^T \overline{\mathbf{d}}}$ with $\lambda_1 = 0$.

9. Property $\langle 7 \rangle$ generalizes to several eigenvalues with eigenvalue 0 for disconnected graphs (the proof is left to the reader as an exercise). If a graph has C subgraphs that are intrinsically connected but not mutually, then $\mathbf{L}$ has

C orthogonal eigenvectors with eigenvalue 0. Each of these eigenvectors has identical values within each of the connected subgraphs and possibly different values between subgraphs. Since it is possible to arbitrarily rotate a set of eigenvectors with identical eigenvalue and still get a set of eigenvectors, it is possible to chose the eigenvectors with eigenvalue 0 such that each one has the value 1 within a subgraph and value 0 on all other nodes. Such vectors are referred to as *indicator vectors* [7]. These indicator vectors can then be normalized to fulfill the convention of normalized eigenvectors.

10. If we do not perform the rotation mentioned in Property $\langle 9 \rangle$ to get indicator vectors, but rather choose the first eigenvector to be the one-vector, then all higher eigenvectors of the ordinary eigenvalue equation $\mathbf{L}\mathbf{u}_\alpha = \gamma_\alpha \mathbf{u}_\alpha$ have zero mean, since $\forall \alpha \neq 1 : 0 \overset{(2)}{=} \mathbf{u}_1^T \mathbf{u}_\alpha \overset{(7)}{\Longleftrightarrow} 0 = \mathbf{1}^T \mathbf{u}_\alpha = \sum_j u_{\alpha j}$ by Properties $\langle 2,7 \rangle$.

11. Similarly, if the first eigenvector is the one-vector all higher eigenvectors of the generalized eigenvalue equation $\mathbf{L}\mathbf{w}_\alpha = \lambda_\alpha \mathbf{D}\mathbf{w}_\alpha$ have weighted zero mean since $\forall \alpha \neq 1 : 0 \overset{(6)}{=} \mathbf{w}_1^T \mathbf{D}\mathbf{w}_\alpha \overset{(7)}{\Longleftrightarrow} 0 = \mathbf{1}^T \mathbf{D}\mathbf{w}_\alpha = \sum_j w_{\alpha j} D_{jj}$ by Properties $\langle 6,7 \rangle$.

12. The eigenvectors are solutions to the optimization problems and the eigenvalues are the values that the objective functions assume for the optimal solutions, see Sect. 2.4. Equation (45) yields $\mathbf{u}_\alpha^T \mathbf{L}\mathbf{u}_\alpha = \gamma_\alpha$, and $\hat{\mathbf{w}}_\alpha^T \hat{\mathbf{L}} \hat{\mathbf{w}}_\alpha = \lambda_\alpha$ holds analogously. For the generalized eigenvalue equation, we find (50) $\mathbf{w}_\alpha^T \mathbf{L}\mathbf{w}_\alpha = \lambda_\alpha$.

Further reading: [9].

3 Algorithms

3.1 Similarity Graphs

The algorithms presented in the following are all based on the properties of the Laplacian matrix discussed above. In order to take advantage of the Laplacian matrix, though, any input data first has to be represented as a graph, commonly referred to as a *similarity graph*: A simple graph where the nodes represent individual data samples and edge weights denote the similarity (or distance) between two connected nodes, i.e. data samples. Appropriate similarity metrics depend on the problem and can be as simple as the Euclidean or Manhattan distance between two points.

Thus, the first step in each algorithm is to construct the similarity graph, which is described here. There are different ways to construct a similarity graph, depending on the problem at hand [1,2,6]. Three common methods are ϵ-neighborhood, k-nearest neighbors, and fully connected graphs:

- **ϵ-neighborhood:** Two nodes are connected if the distance between them is smaller than a given threshold ϵ. Often ϵ is chosen so small that the distance values within an ϵ-neighborhood do not carry much useful information. In this case edges are often weighted binary, i.e., with 1 or 0 depending on whether the data samples in question are close enough or not, respectively.

Table 1. Overview over different Laplacian matrices, eigenvalue equations, optimization problems, and solutions.

	Weight matrix: $\mathbf{W}$				
(29)	Degree matrix: $\mathbf{D} : D_{ij} := \delta_{ij} \sum_j W_{ij}$			(56)	$\underline{\mathbf{D}} := \mathrm{diag}(1/\sqrt{D_{11}}, ..., 1/\sqrt{D_{II}})$
(32)	Laplacian matrix: $\mathbf{L} = \mathbf{D} - \mathbf{W}$			(63)	Sym. norm. Lapl. matrix: $\hat{\mathbf{L}} := \underline{\mathbf{D}}\mathbf{L}\underline{\mathbf{D}}$
⟨1⟩	Is symmetric: $\mathbf{L} = \mathbf{L}^T$			⟨1⟩	Is symmetric: $\hat{\mathbf{L}} = \hat{\mathbf{L}}^T$
⟨3⟩	Is positive semi-definite: $\mathbf{x}^T\mathbf{L}\mathbf{x} \geq 0 \ \forall \mathbf{x}$			⟨3⟩	Is positive semi-definite: $\mathbf{x}^T\hat{\mathbf{L}}\mathbf{x} \geq 0 \ \forall \mathbf{x}$
(41)	Ordinary eigenvalue equation: $\mathbf{L}\mathbf{u}_\alpha = \gamma_\alpha \mathbf{u}_\alpha$	(46)	Generalized eigenvalue equation: $\mathbf{L}\mathbf{w}_\alpha = \lambda_\alpha \mathbf{D}\mathbf{w}_\alpha$	(60)	Ordinary eigenvalue equation: $\hat{\mathbf{L}}\hat{\mathbf{w}}_\alpha = \lambda_\alpha \hat{\mathbf{w}}_\alpha$
	Optimization problem: minimize		Optimization problem: minimize		Optimization problem: minimize
(42)	$\mathbf{u}_\alpha^T\mathbf{L}\mathbf{u}_\alpha = \frac{1}{2}\sum_{ij}(u_{\alpha,i} - u_{\alpha,j})^2 W_{ij}$	(47)	$\mathbf{w}_\alpha^T\mathbf{L}\mathbf{w}_\alpha = \frac{1}{2}\sum_{ij}(w_{\alpha,i} - w_{\alpha,j})^2 W_{ij}$	(68)	$\hat{\mathbf{w}}_\alpha^T\hat{\mathbf{L}}\hat{\mathbf{w}}_\alpha = \frac{1}{2}\sum_{ij}\left(\frac{\hat{w}_{\alpha,i}}{\sqrt{D_{ii}}} - \frac{\hat{w}_{\alpha,j}}{\sqrt{D_{jj}}}\right)^2 W_{ij}$
(43,44)	subject to $\mathbf{u}_\beta^T\mathbf{u}_\alpha = \delta_{\beta\alpha} \ \forall \beta \leq \alpha$	(48,49)	subject to $\mathbf{w}_\beta^T\mathbf{D}\mathbf{w}_\alpha = \delta_{\beta\alpha} \ \forall \beta \leq \alpha$		subject to $\hat{\mathbf{w}}_\beta^T\hat{\mathbf{w}}_\alpha = \delta_{\beta\alpha} \ \forall \beta \leq \alpha$
	Trivial first solution:		Trivial first solution:		Trivial first solution:
⟨7⟩	$\mathbf{u}_1 = \mathbf{1}/\sqrt{\mathbf{1}^T\mathbf{1}}$ with $\gamma_1 = 0$	⟨7⟩	$\mathbf{w}_1 = \mathbf{1}/\sqrt{\mathbf{1}^T\mathbf{D}\mathbf{1}}$ with $\lambda_1 = 0$	⟨8⟩	$\hat{\mathbf{w}}_1 = \overline{\mathbf{d}}/\sqrt{\overline{\mathbf{d}}^T\overline{\mathbf{d}}}$ with $\lambda_1 = 0$
	Objective function value:		Objective function value:		Objective function value:
⟨12⟩,(45)	$\mathbf{u}_\alpha^T\mathbf{L}\mathbf{u}_\alpha = \gamma_\alpha$	⟨12⟩,(50)	$\mathbf{w}_\alpha^T\mathbf{L}\mathbf{w}_\alpha = \lambda_\alpha$	⟨12⟩	$\hat{\mathbf{w}}_\alpha^T\hat{\mathbf{L}}\hat{\mathbf{w}}_\alpha = \lambda_\alpha$
			Relation between two solutions:		
		⟨5⟩,(57,60)	$\lambda_\alpha = \lambda_\alpha$		
		⟨5⟩,(62)	$\mathbf{w}_\alpha = \underline{\mathbf{D}}\hat{\mathbf{w}}_\alpha$		

- k-**nearest neighbors:** Node v_i is connected to v_j if v_j is among the k nearest neighbors of v_i. Note that this neighborhood relation is not symmetric and yields a directed graph, thus some cleanup is required. To arrive at a simple graph we take each unilateral edge that has no mirrored counterpart and either remove it or keep it and set it as bilateral. Removal results in a graph where each node has at most k neighbors (*mutual k-nearest neighbor graph*), while setting unilateral edges to bilateral results in a graph where each node has at least k neighbors (*k-nearest neighbor graph*). All edges are weighted by the similarity between the two nodes they connect. Binary weighting, as in the preceding method, is more dangerous here, because it cannot be guaranteed that connected nodes are close to each other.
- **Fully connected:** To construct a fully connected graph each data sample is simply connected to all others. In this case, using binary weights renders the graph entirely meaningless. A fully connected graph always requires weighting the edges with a similarity function (e.g. a Gaussian similarity function for vectorial data $w_{ij} = w_{ji} = s(\mathbf{x}_i, \mathbf{x}_j) = \exp(-||\mathbf{x}_i - \mathbf{x}_j||^2/(2\sigma^2))$ where σ defines the extent of local neighborhoods).

3.2 Laplacian Eigenmaps (LEM)

Motivation. Many algorithms work only on vectorial data and are limited in the dimensionality they can process efficiently. This causes problems if one has data that is either not vectorial, such as text, or too high dimensional, such as images, or both. If one can define a similarity function on the data, yielding a scalar similarity value for each pair of data samples, the Laplacian eigenmaps algorithm can provide a low-dimensional vectorial embedding of the data that tends to preserve similarity relationships and allows to apply other algorithms to the data that would not be applicable directly [1]. Laplacian eigenmaps are also very good for a 2- or 3-dimensional visualization of data.

For example, imagine a drone hovering through the air while equipped with a downward facing camera. Using the high dimensional pictures from its camera, we could, in theory, precisely compute the drone's current position and elevation. Unfortunately, the space of all possible high dimensional images is effectively intractable. Luckily though, we are merely interested in a small subset of this space, namely only those images the drone's camera can actually produce in a particular environment. And while each data point of this vastly smaller subset still is of the original, high dimensionality, it can be fully described by six dimensions alone: the position and orientation of the drone in 3D space. Laplacian eigenmaps can be used to find a low dimensional embedding of the images that still permits extracting positional and orientation information.

Objective. The objective of the Laplacian eigenmaps algorithm is to find an embedding of a set of I data samples (do not need to be vectors, but there must be a similarity function) in a low-dimensional vector space $\{\mathbf{y}_1, ... \mathbf{y}_I\}$ such that samples with high similarity are close to each other in the embedding. For dimensionality $M = 1$, i.e. an embedding in only a 1-dimensional space, this objective translates into minimizing

$$\frac{1}{2} \sum_{ij} (y_i - y_j)^2 W_{ij} \tag{73}$$

where the y_i are the values assigned to the samples and W_{ij} indicates the similarity between two samples. We have already seen above how this optimization problem is solved by the second eigenvector of the Laplacian matrix, (42) or (47) depending on the constraint. Each additional eigenvector adds one orthogonal (meaning the values are uncorrelated) dimension to the embedding provided by the other eigenvectors already. The quality of the embedding induced by each eigenvector is given by its associated eigenvalue, which directly relates to the actual value of sum (73). The best M-dimensional embedding is thus given by the first M eigenvectors $\mathbf{u}_\alpha$ or $\mathbf{w}_\alpha$ of the Laplacian matrix with smallest eigenvalues (excluding the first one).

Please notice that the dimension of the eigenvectors corresponds to the number I of data points, because the Laplacian matrix is $I \times I$ by construction. Thus, if you arrange the first M eigenvectors as rows in a matrix, this matrix will be $M \times I$ and the column vectors are the data points $\mathbf{y}_i$ in the M-dimensional embedding. For instance, three data samples embedded in a 2-dimensional space with LEM using (left) the ordinary eigenvalue problem and (right) the generalized eigenvalue problem could yield

$$\begin{pmatrix} & \mathbf{y}_1 & \mathbf{y}_2 & \mathbf{y}_3 \\ & \downarrow & \downarrow & \downarrow \\ \mathbf{u}_2 \rightarrow & -1/\sqrt{2} & 0 & +1/\sqrt{2} \\ \mathbf{u}_3 \rightarrow & -1/\sqrt{6} & +2/\sqrt{6} & -1/\sqrt{6} \end{pmatrix} \quad \begin{pmatrix} & \mathbf{y}_1 & \mathbf{y}_2 & \mathbf{y}_3 \\ & \downarrow & \downarrow & \downarrow \\ \mathbf{w}_2 \rightarrow & w_{2,1} & w_{2,2} & w_{2,3} \\ \mathbf{w}_3 \rightarrow & w_{3,1} & w_{3,2} & w_{3,3} \end{pmatrix} \tag{74}$$

As usual, we have dropped $\mathbf{u}_1$ and $\mathbf{w}_1$, because they have equal components throughout, e.g. $\mathbf{u}_1 = (1,1,1)^T/\sqrt{3}$; $\mathbf{u}_2$ and $\mathbf{u}_3$ have zero mean, because they need to be orthogonal to $\mathbf{u}_1$; and $\mathbf{u}_2$ and $\mathbf{u}_3$ are orthogonal to each other as well. Analogous relations hold for $\mathbf{w}_\alpha$, but are numerically less intuitive.

We now have all the required components to formulate the Laplacian eigenmaps algorithm with constraints (48, 49).

Algorithm

Laplacian Eigenmaps Algorithm [1]

1. Given a set of I data samples, construct a similarity graph G according to one of the methods described in Sect. 3.1.

2. Construct the $I \times I$ weight matrix $\mathbf{W}$, degree matrix $\mathbf{D}$ (29), and Laplacian matrix $\mathbf{L}$ (32) for G as described in Sect. 2.2.
3. Compute the first $M+1$ eigenvectors $\mathbf{w}_\alpha$ of the generalized eigenvalue problem

$$\mathbf{L}\mathbf{w}_\alpha = \lambda_\alpha \mathbf{D}\mathbf{w}_\alpha \qquad (75)$$

ordered by increasing eigenvalues, see Sect. 2.4.
4. An M-dimensional representation $\mathbf{y}_i$ of data sample i is now given by $(w_{2,i}, ..., w_{M+1,i})^T$, see (74).

Sample Application. Figures 6 and 7 show an application of Laplacian eigenmaps to a set of 300 frequently used words [1]. Each word was represented by a 600-dimensional vector indicating how often any of the other words was found to the left or to the right of the considered word. Similarity was defined based on these 600-dimensional vectors. Zooming into Fig. 7 shows that grammatically closely related words are grouped together.

Further reading: [1].

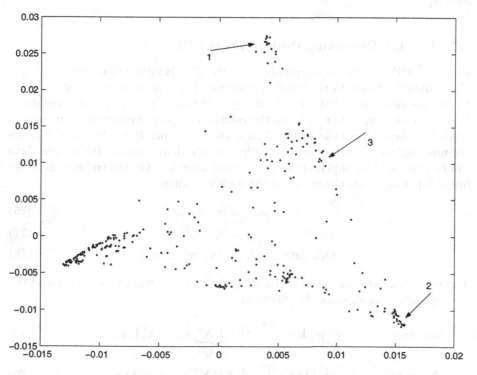

Fig. 6. Dimensionality reduction for 300 frequently used words from their word context data. Figure by Belkin and Niyogi, 2003 [1].

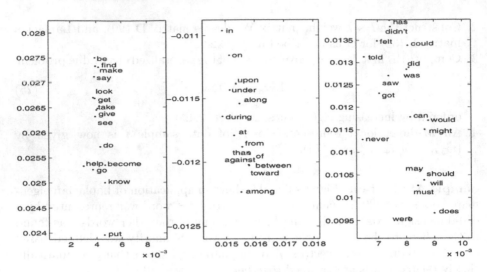

Fig. 7. Zoom-in into the three subregions marked in Fig. 6. Left infinitives, middle prepositions, and right mostly modal and auxiliary verbs. Figure by Belkin and Niyogi, 2003 [1].

3.3   Locality Preserving Projections (LPP)

Linear LPP. Laplacian eigenmaps have the disadvantage that they only provide values for the data used during training. There is no straight forward way to process new data. This can be changed if the nodes v_i are data points in Euclidean space $v_i = \mathbf{x}_i \in \mathbb{R}^N$ and the values of the eigenvectors $\mathbf{w}_\alpha$ are approximated by linear functions in the data points [2]. Since the values of the nodes are now computed with a linear function rather than assigned freely, new data can be processed by applying the same linear function. On the training data the linear function yields the values of the nodes as follows

$$w_{\alpha,i} = \mathbf{x}_i^T \mathbf{z}_\alpha \tag{76}$$

$$\Longleftrightarrow \qquad \mathbf{w}_\alpha = \mathbf{X}^T \mathbf{z}_\alpha \tag{77}$$

$$\text{with data} \qquad \mathbf{X} := (\mathbf{x}_1, \mathbf{x}_2, ..., \mathbf{x}_I) \tag{78}$$

The vectors $\mathbf{z}_\alpha$ are the variables to be optimized. Inserting this in (47) and the corresponding constraints (48, 49) yields

$$\text{minimize} \qquad \mathbf{w}_\alpha^T \mathbf{L} \mathbf{w}_\alpha \overset{(77)}{=} \mathbf{z}_\alpha^T \underbrace{\mathbf{X} \mathbf{L} \mathbf{X}^T}_{=: \mathbf{L}'} \mathbf{z}_\alpha = \mathbf{z}_\alpha^T \mathbf{L}' \mathbf{z}_\alpha \tag{79}$$

$$\text{subject to} \qquad 1 = \mathbf{w}_\alpha^T \mathbf{D} \mathbf{w}_\alpha \overset{(77)}{=} \mathbf{z}_\alpha^T \underbrace{\mathbf{X} \mathbf{D} \mathbf{X}^T}_{=: \mathbf{D}'} \mathbf{z}_\alpha = \mathbf{z}_\alpha^T \mathbf{D}' \mathbf{z}_\alpha \tag{80}$$

$$\text{and} \qquad 0 = \mathbf{w}_\beta^T \mathbf{D} \mathbf{w}_\alpha \overset{(77)}{=} \mathbf{z}_\beta^T \underbrace{\mathbf{X} \mathbf{D} \mathbf{X}^T}_{=: \mathbf{D}'} \mathbf{z}_\alpha = \mathbf{z}_\beta^T \mathbf{D}' \mathbf{z}_\alpha \quad \forall \beta < \alpha \tag{81}$$

This optimization problem can again be solved through a generalized eigenvalue problem, much like the original one. Notice, however, that the eigenvalues and the approximated eigenvectors $\mathbf{w}_\alpha$ are not necessarily identical to those of the original eigenvalue problem, because $\mathbf{w}_\alpha \in \mathbb{R}^I$ is not free but constrained to be a linear function in the $\mathbf{x}_i \in \mathbb{R}^N$. Notice also that this problem is not of the dimensionality of the number I of data points as before but only of the dimension N of the data points, which is usually much smaller and, consequently, makes this approximation more computationally efficient. For instance, if you have 100 data points in 3D, the problem is 3-dimensional not 100-dimensional as for the LEM algorithm. The main advantage, however, is that new data points $\mathbf{x}_j$ can easily be mapped into the low-dimensional space by applying the linear function $\mathbf{x}_j^T \mathbf{z}_\alpha$. Performing Laplacian eigenmaps with this linear approximation is referred to as *locality preserving projections* (LPP).

Sample Application. An application of LPP to face images of a single person is shown in Fig. 8 [2]. Even though the mapping is only linear, LPP still captures some prominent variations and orders the images nicely in 2D. The person looks to the left (or right) at the top (or bottom) of the plot, and it smiles on the right side while it makes faces on the left.

Nonlinear LPP. LPP can be generalized to nonlinear functions by adding a nonlinear expansion prior to the algorithm. Assume $\mathbf{f}(\mathbf{x})$ is such a nonlinear expansion from $\mathbb{R}^N \to \mathbb{R}^P$ with $N \ll P$, then one can define

$$w_{\alpha,i} = \mathbf{f}(\mathbf{x}_i)^T \mathbf{z}_\alpha \tag{82}$$

$$\Longleftrightarrow \quad \mathbf{w}_\alpha = \mathbf{F}^T \mathbf{z}_\alpha \tag{83}$$

$$\text{with} \quad \mathbf{F} := (\mathbf{f}(\mathbf{x}_1), \mathbf{f}(\mathbf{x}_2), ..., \mathbf{f}(\mathbf{x}_I)) \tag{84}$$

and then run the algorithm as before. Notice that now $\mathbf{z}_\alpha \in \mathbb{R}^P$ rather than $\mathbb{R}^N$.
  Further reading: [2].

3.4   Spectral Clustering

Objective. *Spectral clustering* is an umbrella term for a number of algorithms that use the eigenvectors of the Laplacian matrix to perform clustering on a given set of data points. In particular, spectral clustering is often used in image processing to identify connected parts of a given image and, ideally, identify the extent of the individual components of an image, a process called *image segmentation.*

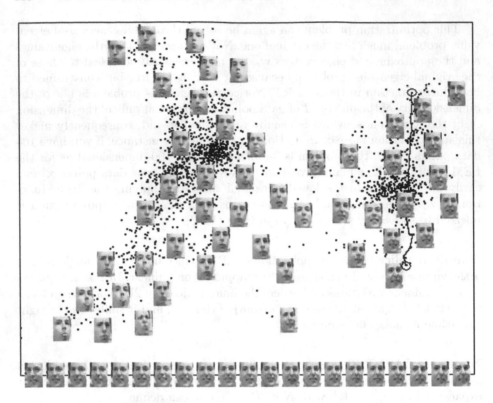

Fig. 8. Dimensionality reduction of face images of a single person down to two dimensions with linear LPP. Face images in the plot indicate what some points stand for and the line of faces at the bottom corresponds to the line of data points on the right. Figure by He and Niyogi, 2004 [2].

As illustrated intuitively in Fig. 2 the eigenvectors of the Laplacian matrix place the nodes of connected subgraphs at the same location, even in two, three, or higher dimensions, if the graph has several subgraphs. This also holds for the eigenvectors of the generalized eigenvalue problem, and this also holds approximately if the subgraphs are not completely separate from each other. Given this representation it is much easier than on the original data to cluster the nodes with some standard clustering algorithm.

Remember that for C intrinsically connected but mutually disconnected subgraphs, i.e. clusters, there are exactly C eigenvectors with constant values on each of the clusters. For extracting C clusters one would therefore use the first C eigenvectors, this time including also the first one, see Property ⟨9⟩.

Algorithm

Normalized Spectral Clustering Algorithm [3]

1. Given a set of I data samples, construct a similarity graph G according to one of the methods described in Sect. 3.1. For instance, when performing segmentation on a single image, each pixel becomes a node of the graph with similarity between nodes usually being a function of color and spatial distance.
2. Compute the weight matrix $\mathbf{W}$, degree matrix $\mathbf{D}$ (29), and Laplacian matrix $\mathbf{L}$ (32) for G.
3. Compute the first C eigenvectors of the generalized eigenvalue problem

$$\mathbf{L}\mathbf{w}_\alpha = \lambda_\alpha \mathbf{D}\mathbf{w}_\alpha \tag{85}$$

   ordered by increasing eigenvalue.
4. Arrange the eigenvectors $\mathbf{w}_1, .., \mathbf{w}_C$ in the rows[3] of a matrix $\mathbf{U}$ and normalize its columns to one to get matrix $\mathbf{T}$ with

$$T_{ij} = U_{ij} / \left(\sum_{i'} U_{i'j}^2 \right)^{1/2} \tag{86}$$

   A C-dimensional representation $\mathbf{y}_i$ of data sample i is now given by the i-th column vector of $\mathbf{T}$.
5. Perform the k-means algorithm on the set of embedded data points $\{\mathbf{y}_1, ... \mathbf{y}_I\}$ to partition the data into C clusters.

Sample Application. Figure 9 shows an example of applying spectral clustering to an old data set collected by Edgar Anderson [11]. He measured length and width of the sepal and petal from 50 exemplars of three types of iris. One species (red in the left plot) is well separated from the other two, which in turn are hard to distinguish in the 2D plots. Spectral clustering performs fairly well on this task in 4D as one can see by comparing ground truth on the left with the clustering result on the right.

Further reading: [6], an excellent tutorial on spectral clustering.

---

[3] In the original formulation [3], the vectors were arranged in columns. We use rows here for consistency with the LEM algorithm, see Sect. 3.2.

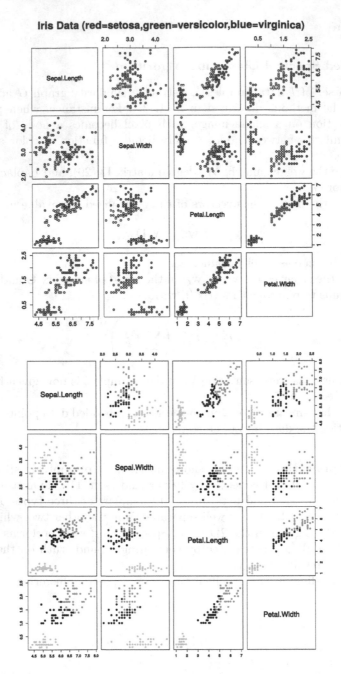

Fig. 9. Spectral clustering on iris (the plant, not the eye) data. Top: length and width of the sepal and petal from 50 exemplars of three types of iris as indicated by the three colors. Figure by Nicoguaro on Wikimedia, 2017 [4]. Bottom: Result of spectral clustering on the 150 four-dimensional data points. Figure by Sigbert on Wikimedia, 2017 [5].

Acknowledgments. We thank Jan Melchior and Merlin Schüler for valuable feedback on an earlier version of these lecture notes.

References

1. Belkin, M., Niyogi, P.: Laplacian eigenmaps for dimensionality reduction and data representation. Neural Comput. **15**(6), 1373–1396 (2003)
2. He, X., Niyogi, P.: Locality preserving projections. In: Thrun, S., Saul, L., Schölkopf, B. (eds.) Advances in Neural Information Processing Systems, pp. 153–160. MIT Press (2004). http://papers.nips.cc/paper/2359-locality-preserving-projections.pdf
3. Ng, A.Y., Jordan, M.I., Weiss, Y.: On spectral clustering: analysis and an algorithm. In: Advances in Neural Information Processing Systems, vol. 14, pp. 849–856 (2002)
4. Nicoguaro on Wikimedia Commons: File:Iris dataset scatterplot.svg – wikimedia commons, the free media repository (2017). https://commons.wikimedia.org/w/index.php?title=File:Iris_dataset_scatterplot.svg&oldid=235116001. Accessed 2 Dec 2017
5. Sigbert on Wikimedia Commons: File:specclus iriscluster.svg – wikimedia commons, the free media repository (2017). https://commons.wikimedia.org/w/index.php?title=File:Specclus_iriscluster.svg&oldid=235116126. Accessed 2 Dec 2017
6. Von Luxburg, U.: A tutorial on spectral clustering. Stat. Comput. **17**(4), 395–416 (2007)
7. Wikipedia: Indicator vector – Wikipedia, The Free Encyclopedia (2016). https://en.wikipedia.org/w/index.php?title=Indicator_vector&oldid=743797854. Accessed 3 Dec 2017
8. Wikipedia: Graph (discrete mathematics) – Wikipedia, The Free Encyclopedia (2017). https://en.wikipedia.org/w/index.php?title=Graph_(discrete_mathematics)&oldid=800782160. Accessed 3 Dec 2017
9. Wikipedia: Laplacian matrix – Wikipedia, The Free Encyclopedia (2017). https://en.wikipedia.org/w/index.php?title=Laplacian_matrix&oldid=812863352. Accessed 3 Dec 2017
10. Wikipedia: Rayleigh quotient – Wikipedia, The Free Encyclopedia (2017). https://en.wikipedia.org/w/index.php?title=Rayleigh_quotient&oldid=808561799. Accessed 3 Dec 2017
11. Wikipedia: Spektrales Clustering – Wikipedia, Die freie Enzyklopädie (2017). https://de.wikipedia.org/w/index.php?title=Spectral_Clustering&oldid=170428156. Accessed 2 Dec 2017
12. Wikipedia: Stochastic matrix – Wikipedia, The Free Encyclopedia (2017). https://en.wikipedia.org/w/index.php?title=Stochastic_matrix&oldid=813141273. Accessed 3 Dec 2017
13. Wiskott, L., Schönfeld, F.: Laplacian matrix for dimensionality reduction and clustering - lecture notes. e-print arXiv:1909.08381 (2019)

Author Index

Author Index

Printed in the United States
By Bookmasters